COLLEGE MATHEMATICS USING

THE TEXAS INSTRUMENTS GRAPHING CALCULATOR

SEVENTH EDITION

COLLEGE MATHEMATICS

FOR BUSINESS, ECONOMICS, LIFE SCIENCES,

AND SOCIAL SCIENCES

CAROLYN L. MEITLER
CONCORDIA UNIVERSITY, WISCONSIN

COLLEGE MATHEMATICS USING
THE TEXAS INSTRUMENTS GRAPHING CALCULATOR

SEVENTH EDITION

COLLEGE MATHEMATICS

FOR BUSINESS, ECONOMICS, LIFE SCIENCES,
AND SOCIAL SCIENCES

RAYMOND A. BARNETT
MICHAEL R. ZIEGLER

PRENTICE HALL, Upper Saddle River, NJ 07458

Production Editor:*Ann Marie Longobardo*
Production Supervisor*: Joan Eurell*
Acquisitions Editor: *George Lobell*
Supplements Acquisitions Editor: *Audra J. Walsh*
Production Coordinator: *Alan Fischer*

Simon & Schuster / A Viacom Company
Upper Saddle River, New Jersey 07458

Printed in the United States of America

10 9 8 7 6 5 4 3 2 1

ISBN: 0-13-232851-8

Prentice-Hall International (UK) Limited, *London*
Prentice-Hall of Australia Pty. Limited, *Sydney*
Prentice-Hall Canada Inc., *Toronto*
Prentice-Hall Hispanoamericana, S.A., *Mexico*
Prentice-Hall of India Private Limited, *New Delhi*
Prentice-Hall of Japan, Inc., *Tokyo*
Simon & Schuster Asia Pte. Ltd., *Singapore*
Editora Prentice-Hall do Brasil, Ltda., *Rio de Janeiro*

Table of Contents

PREFACE

Graphing calculators are a powerful tool that can help in learning mathematics. This manual includes examples that illustrate how to use the calculator with topics covered in **College Mathematics for Business, Economics, Life Sciences, and Social Sciences**, seventh edition by R. A. Barnett and M. R. Ziegler. A detailed solution to each example shows how a graphing calculator can be used to solve each problem.

Appendices

Three appendices are included. Each appendix contains keystrokes necessary to solve problems for a particular calculator.

Appendix A	TI-81 Operations
Appendix B	TI-82 Operations
Appendix C	TI-85 Operations

Keystrokes necessary to solve a problem are shown. More than one method of solution using the calculator is shown for many examples. Neither programming nor parametric equations are used.

The appendices correspond to one another. The same example is used in each appendix. This allows instruction to be given to a class of students in which more than one calculator model is being used.

Exercise sets

Exercise sets are found at the end of each chapter in the manual. These are intended to give practice on the types of problems discussed in this manual.

Calculations

Default settings on the calculator have been used for all calculations in the problem solutions, unless otherwise noted. Numbers are rounded to the desired number of significant digits or number of decimal places as the problem requires after all calculations have been completed. Directions on how to set the calculator to a specified number of significant digits, or a given number of decimal places, before performing the calculations are given in Appendices A-17, B-17, and C-17. This was not done in the examples of this manual but might be convenient in many problems.

Built-in Calculator Functions
The built-in functions (features) of the calculator are not used in the examples of the chapters. Discussion of some of the built-in features for a particular calculator can be found in the appendix pertaining to that calculator.

Graphs
The curves shown in the figures of this manual are not smooth but rather are intended to represent the graphing calculator display.

ACKNOWLEDGMENTS

The completion of this project would not have occurred without the support and assistance of several individuals. First, and foremost, are the textbook authors Raymond A. Barnett and Michael R. Ziegler. Their suggestions were most helpful in forming the philosophy, the layout, and the content of this manual. Also I would like to thank Suzanne Skony who patiently read the manuscript and worked through all the problems. Her help in checking for accuracy and consistency was invaluable. Finally, I thank my family for their encouragement and support.

Carolyn L. Meitler

Chapter 1

A Beginning Library of Elementary Functions

This chapter contains examples using the calculator to illustrate:

- Plotting points and drawing lines
- Graphing a function
- Evaluating algebraic expressions
- Determining the domain of a function from the graph
- Evaluating a function for a given value of x
- Investigating the change in the value of an expression as the value of a variable changes
- Investigating the change in the graph of a function when constant terms are changed
- Finding intercepts
- Solving an equation in one variable
- Solving an inequality in one variable

The features of a graphing calculator used are:

- Calculation
- Storing an algebraic expression
- Evaluating an algebraic expression
- Graphing and viewing window variables
- Graphing a function
- Zoom
- Trace

EXAMPLE 1 ***TEXTBOOK SECTION 1-1***

(A) Plot the points (-5, 2) and (1, 3). (B) Draw a line between them.

Solution: (A) First the size of the viewing window needs to be set. Set it at Xmin=-10, Xmax=10, Xscl=1, Ymin=-10, Ymax=10, and Yscl=1 (see Appendix Section A-6 & A-7 , B-6 & B-7, or C-6 & C-7 Example 1 Method 1 of this manual). Get the [Pt-On(] function on the calculator. Enter the values -5 and 2 and plot the first point. Get the [Pt-On(] function again and plot the second point. (See Appendix Section A-3, B-3, or C-3 of this manual.)

(B) Get the [Line(] function of the calculator and draw the line. (See Appendix Section A-3, B-3, or C-3 of this manual.)

EXAMPLE 2 ***TEXTBOOK SECTION 1-1***

(A) Find the y coordinate for $y = 3 - x^2$ for $x = -3, -1, 0, 1, 3$. (B) Plot these ordered pairs. (C) Graph the function.

Solution: (A) Store $3 - x^2$ in the function list in your calculator as: 3 - X ^ 2. (See Appendix Section A-4, B-4, or C-4 of this manual on how to store an expression and evaluate it.) Store the first value for x in the calculator and evaluate the function. Repeat for the other 4 values of x. The resulting ordered pairs are (-3, -6), (-1, 2), (0, 3), (1, 2), and (3, -6).

(B) Refer to Example 1 above on how to plot points on the calculator.

(C) Set the viewing window variables to [-10,10]1 by [-10, 10]1 (see Appendix Section A-6 & A-7 , B-6 & B-7, or C-6 & C-7 Example 1 Method 1 of this manual). Store $3 - x^2$ in the function list in your calculator if it has not already been stored.

Clear or deselect all other functions. Clear all drawings. (See Appendix Section A-6, B-6, or C-6 Example 1 of this manual.)

If you are using a TI-81 or TI-85 clear all statistical registers. (See Appendix Section A-16 or B-16 of this manual.) If you are using a TI-82, turn off all plots on the STAT PLOT menu.

Graph the function (see Appendix Section A-6, B-6 or C-6 of this manual).

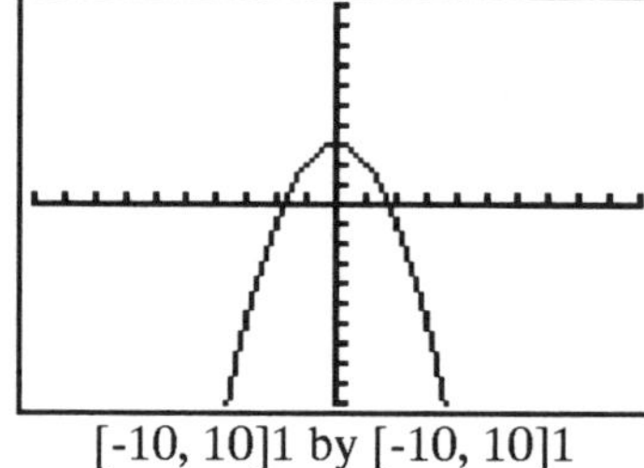
[-10, 10]1 by [-10, 10]1

EXAMPLE 3 ***TEXTBOOK SECTION 1-1***

Refer to Example 3 of Textbook Section 1-1. Graph $y = \sqrt{4-x}$ and determine the domain by examining the graph.

Solution: Store the function in the function list in the calculator as: √ (4 - X). Set the viewing window variables to [-5,5]1 by [-5, 5]1 (see Appendix Section A-6, B-6, or C-6 Example 1 of this manual). Graph the function. Use TRACE and ZOOM to find the x coordinate of the right-hand end of the graph. We see that this is 4. Evaluating the function at this value, we find that y=0. Hence the domain of the function is $\{x \mid x \leq 4\}$.

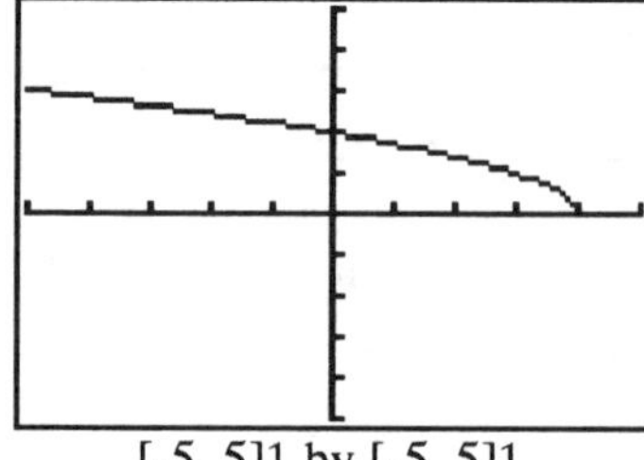
[-5, 5]1 by [-5, 5]1

EXAMPLE 4 — *TEXTBOOK SECTION 1-1*

Let $f(x) = 2x^2+3$. (A) Fill in the chart below. (B) How does the quantity $\frac{f(2+h) - f(2)}{h}$ change as h assumes values .2, .1, .01, and .001. (C) Make a conjecture as to what happens to this quantity when h gets closer and closer to zero.

h	.2	.1	.01	.001	→	0
$\frac{f(2+h) - f(2)}{h}$						

Solution: (A) $f(2+h) = 2(2+h)^2+3$ and $f(2) = 11$. Store $2(2+h)^2+3$ as 2 (2 + [ALPHA] H) ^ 2 + 3 as Y1 and 11 as Y2 . Store (Y1-Y2)÷H as Y3 by entering:

For the TI-81 and TI-82

[(] [Y-VARS] [1] <Y1> [-] [Y-VARS] [2] <Y2> [)] [÷] [ALPHA] [H]

For the TI-85

[(] [2nd] [ALPHA] [Y] [1] [-] [2nd] [ALPHA] [Y] [2] [)] [÷] [ALPHA] [H]

Store .2 as H and evaluate Y3 by recalling it (see Appendix Section A-4, B-4, or C-4 of this manual). Repeat by storing each of the other values of h and evaluating Y3. The results are shown below.

h	.2	.1	.01	.001	→	0
$\frac{f(2+h) - f(2)}{h}$	8.4	8.2	8.02	8.002	→	8

(B) & (C) The quantity gets closer and closer to 8 as h gets closer and closer to 0.

EXAMPLE 5 — *TEXTBOOK SECTION 1-1*

Refer to Example 7 of Textbook Section 1-1. The price-demand function that models the data is $p(x)=94.8 - 5x$.

(A) Evaluate the price-demand function at $x = 2$, 5, 8, and 12 million.

(B) Find the error between the result of using the function to calculate the price and the actual data shown in Table 4 of Textbook Section 1-1.

(C) Plot the points and graph the price-demand function that models the data.

Solution: (A) Store the function $p(x)=94.8 - 5x$ as Y1 as 94.8 - 5 X. Store 2 as X. Recall the function Y1 and evaluate. (See Appendix Section A-4, B-4, or C-4 of this manual). The results are shown in the first two columns of the table to the right.

x	$p(x)$	p	Error $p(x)$-p
2	84.8	87	-2.2
5	69.8	68	1.8
8	54.8	53	1.8
12	34.8	37	-2.2

(B) Calculate $p(x)$-p to find the error. The results are shown in the table on the preceeding page.

(C) Set the viewing window variables at [0, 15]1 by [0, 100]10. (See Appendix Section A-8, B-8, or C-8 of this manual on how to choose and set the viewing window variables.) Plot the points (see Example 1 above). Store the price-demand function as Y1 as 94.8 - 5 X and graph.

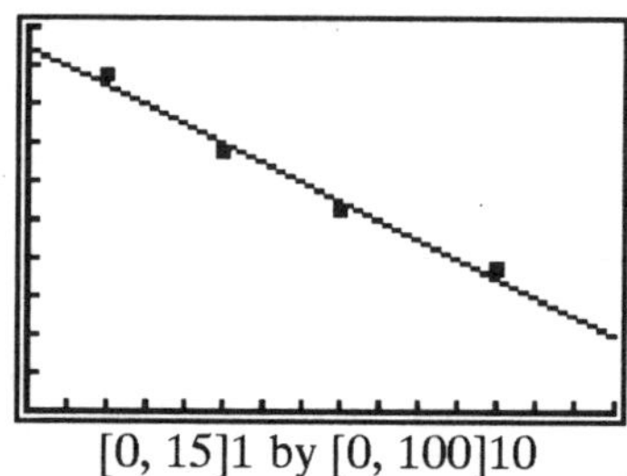
[0, 15]1 by [0, 100]10

EXAMPLE 6 — TEXTBOOK SECTIONS 1-2 & 1-4

Graph $f(x) = x^2$, $f(x) = x^2+4$, and $f(x) = x^2 - 7$ on the same coordinate axes. Compare the graphs and describe the effect of adding or subtracting a constant from $f(x)=x^2$.

Solution: Store the functions as three separate functions in the calculator as: X^2, X^2+4, and X^2-7. Graph using [-15, 15]1 by [-20, 20]1. We see that the second graph has the same shape but is shifted vertically upwards 4 units from $f(x) = x^2$. The third graph has the same shape but is shifted vertically downwards 7 units from $f(x) = x^2$.

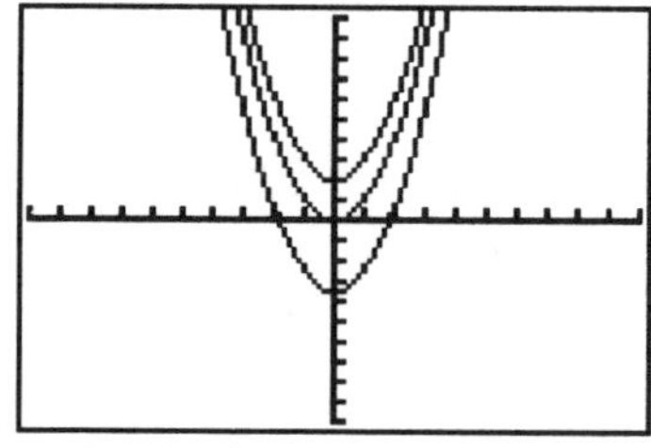
[-15, 15]1.5 by [-20, 20]2

EXAMPLE 7 — TEXTBOOK SECTIONS 1-2 & 1-3

Consider the function $f(x) = |x|-5$. (A) Find the value(s) of x for which $f(x)$ is positive. (B) Discuss how the graph of this function differs from $y = |x|$.

Solution: (A) Store $f(x)$ as Y1 as ABS (X) - 5 where ABS is the built-in absolute value function in your calculator (see Appendix Section A-2, B-2, or C-2 Example 1 of this manual). Set the viewing window at [-10, 10]1 by [-10, 10]1. Graph the function. Use ZOOM and TRACE to find the x intercepts. These are -5 and 5. The x intercepts are the cut-off values for the set of values for which $f(x)$ is positive.

Examine the graph to see where the function is above the y axis. This is where the function is positive. The function is positive for x values to the right of 5 and for x values to the left of -5. Hence solution is the set of all real numbers such that $x \leq -5$ or $x \geq 5$.

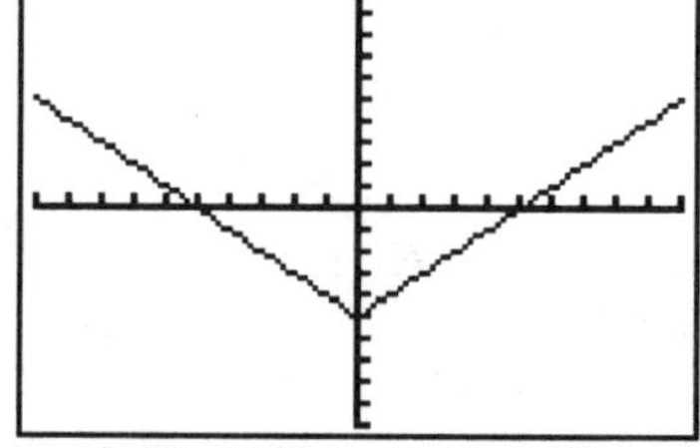
[-10, 10]1 by [-10, 10]1

(B) The graph of $f(x)$ differs from $y=|x|$ in that the graph shifted 5 units downward from that of y = $|x|$.

EXAMPLE 8 ***TEXTBOOK SECTION 1-3***

Use graphing techniques to explore whether or not the two lines $3.15x + 2.79y = 5.38$ and $3.05x + 2.70y = 5.30$ are parallel.

Solution: Graph both lines on the same set of coordinate axes using [-10,10]1 by [-10,10]1. At first glance there appears to be only one line. However, use either the zoom box, zoom in, or set the graph screen at [.5,1.5]0 by [.5,3]0 to see that there are in fact two lines. The lines appear to be parallel. However, their slopes vary slightly and hence are actually two intersecting lines. (The point of intersection is $(58, -63.\bar{5})$).

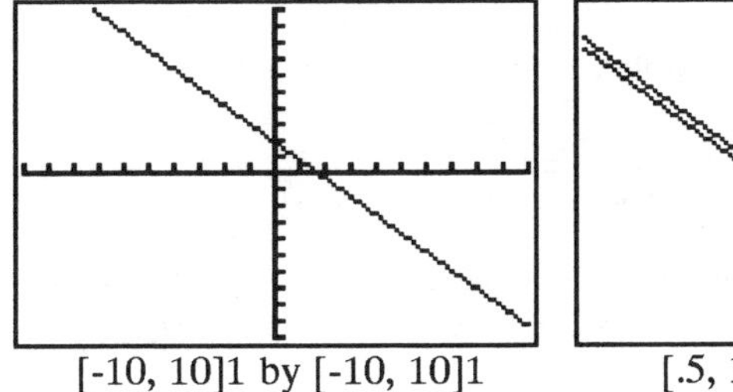

[-10, 10]1 by [-10, 10]1 [.5, 1.5]0 by [.5, 1.5]0

EXAMPLE 9 ***TEXTBOOK SECTION 1-3***

Refer to Explore-Discuss 3 in Textbook Section 1-3.

(A) Graph the five lines for Part (A) and discuss the significance of the constant term.

(B) Graph the five lines for Part (B) and discuss the significance of the slope.

Solution: (A) Store the functions in the function list (only four can be stored on the TI-81). Use [-10, 10]1 by [-10, 10]1 as the viewing window variables (see Appendix Section A-6, B-6, or C-6 of this manual). Graph. By examining the graph we see that the lines are parallel. The slope of the line is 1 (the coefficient on x). The lines have y intercept equal to the constant term.

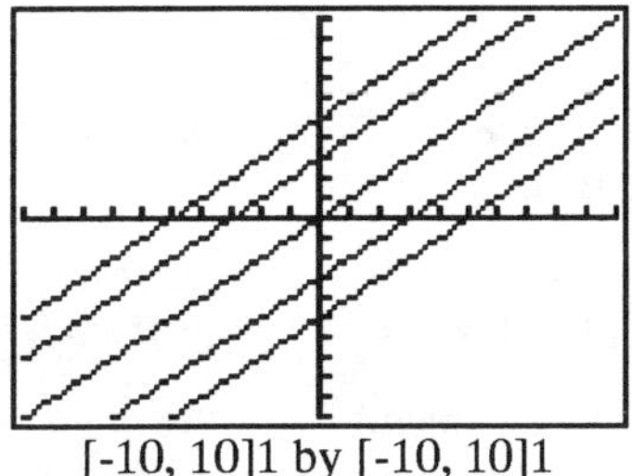

[-10, 10]1 by [-10, 10]1

(B) Store the functions in the function list (only four can be stored on the TI-81). Use [-10, 10]1 by [-10, 10]1 as the viewing window variables (see Appendix Section A-6, B-6, or C-6 of this manual). Graph. By examining the graph we see that the lines all have the same y intercept of -1 but have different slopes. The coefficient on x is the slope of the line.

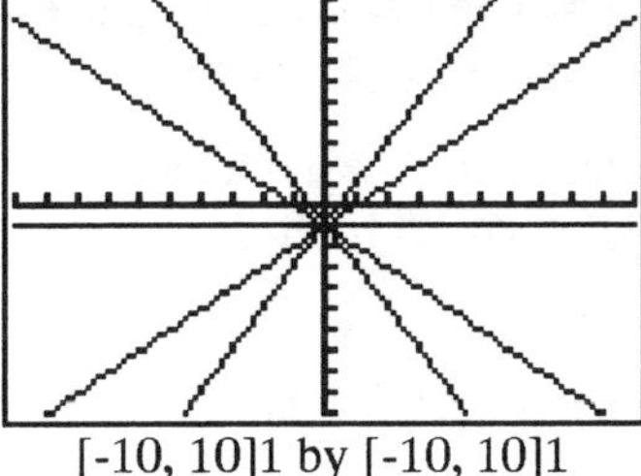

[-10, 10]1 by [-10, 10]1

EXAMPLE 10 ***TEXTBOOK SECTION 1-4***

Refer to Textbook Section 1-4 Example 1.

(A) Graph the function given in this problem, $f(x) = -x^2+5x+3$ and complete Parts (B)-(E) of the Example.

(B) Conjecture what you think will happen to the graph if the 3 is replaced by 6 in the function?

(C) Conjecture what you think will happen to the x and y intercepts if the 3 in $f(x)$ is replaced by a 6.

(D) Check your conjecture by graphing $g(x)=-x^2+5x+6$ on the same set of coordinate axes.

(E) Repeat Parts (B)-(C) of this example for $f(x) = -x^2 + 5x - 6$.

Solution (A): Store $f(x)$ as Y1 in the calculator as - X ^ 2 + 5 X + 3. Set the viewing window variables at [-10, 10]1 by [-10, 10]1. Graph.

Solution (B): Let the new function be named $g(x) = -x^2+5x+6$. A good conjecture, based on Textbook Section 1-2, is that the graph of $g(x)$ will be shifted upward 3 units from $f(x)$.

Solution (C): A good conjecture is that the least x intercept for $g(x)$ will be less than the least x intercept for $f(x)$ and that the greatest x intercept for $g(x)$ will be greater than the greatest x intercept for $f(x)$.

Furthermore, the y intercept of $g(x)$ will be greater than the y intercept for $f(x)$.

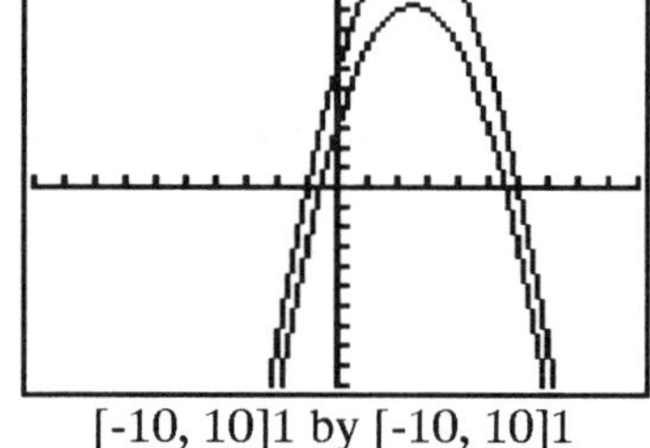
[-10, 10]1 by [-10, 10]1

Solution (D): The graph of $f(x)$ and $g(x)$ are shown to the right. See Appendix Section A-7 & A-9, B-7 & B-9, or C-7 & C-9 on how to use trace and zoom and how to solve equations. See Appendix Section A-4, B-4, or C-4 on using the calculator to evaluate the quadratic formula.

Solution (E): The graph of $h(x) = -x^2+5x-6$ will be shifted downward 9 units from $f(x)$.

A good conjecture for the intercepts is that the least x intercept for $h(x)$ will be greater than the least x intercept for $f(x)$ and that the greatest x intercept for $h(x)$ will be less than the greatest x intercept for $f(x)$ if $h(x)$ intersects the x axis.

Furthermore, the y intercept for $h(x)$ will be less than the y intercept for $f(x)$.

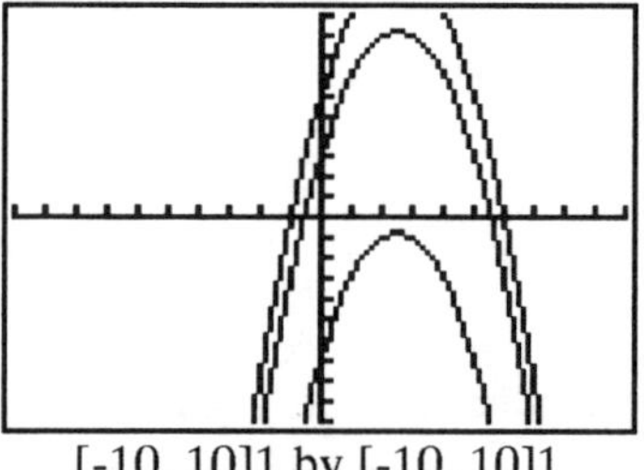
[-10, 10]1 by [-10, 10]1

The graph of $f(x)$, $g(x)$, and $h(x)$ are shown to the right. We see that the graph of $h(x)$ does not intersect the x axis. Hence there are no x intercepts for $h(x)$.

EXAMPLE 11 ***TEXTBOOK SECTION 1-4***

Consider the functions $f(x)=2(x-2)^2+3$ and $g(x)=2(x-4)^2-2$. Discuss the relationship of the graph of $f(x)$ to that of $g(x)$.

Solution: The graph of $g(x)$ will be two units to the left and 5 units downward from $f(x)$. The graphs are shown to the right.

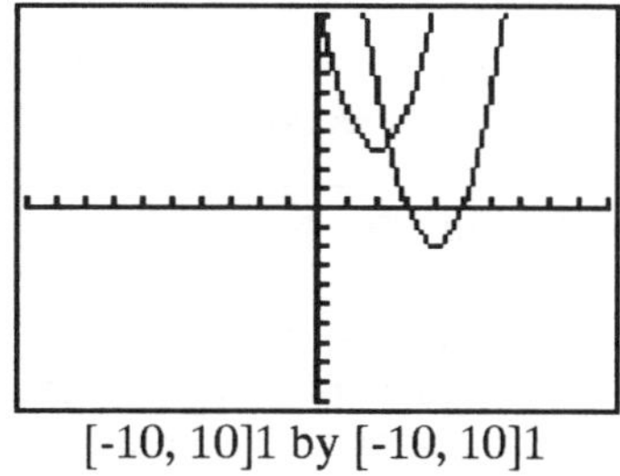

[-10, 10]1 by [-10, 10]1

EXAMPLE 12 ***TEXTBOOK SECTION 1-4***

The market research department of a company recommended to management that the company manufacture and market a promising new product. After extensive surveys, the research department backed up the recommendation with the **price-demand equation** $x = 4600 - 29p$ where x is the number of units that retailers are likely to buy per month at $\$p$ per unit for $0 \le p \le 150$, and where \$71000 is the fixed cost and \$60 is the variable cost per unit.

(A) Find the cost function in terms of the price p.

(B) Find the revenue function in terms of the price p.

(C) Find the price p at which the company will **break-even**. This price is where the cost is equal to the revenue.

Solution (A): The cost function is $C = 71000 + 60x = 71000 + 60(4600 - 29p)$.

Solution (B): The **revenue equation** is $R = xp = (4600 - 29p)p$.

Solution (C): We wish to find the price p at which the cost equals the revenue. Since the calculator uses X as the variable, change p to X and store the cost equation and the revenue equation in the calculator as separate functions as: 71000 + 60 (4600- 29 X) and (4600- 29 X) X .

We determine the appropriate viewing window variables by evaluating the equations for some particular values of X. (See Appendix Sections A-8, B-8, or C-8 of this manual.) The following chart will help.

X	Cost	Revenue
0	341000	0
50	254000	152500
100	167000	160000
150	80000	22500

A first choice for the viewing window variables could be [0,150]10 by [100000,200000]10000. The graph is shown to the right. (See Appendix Section A-8, B-8, or C-8 of this manual.)

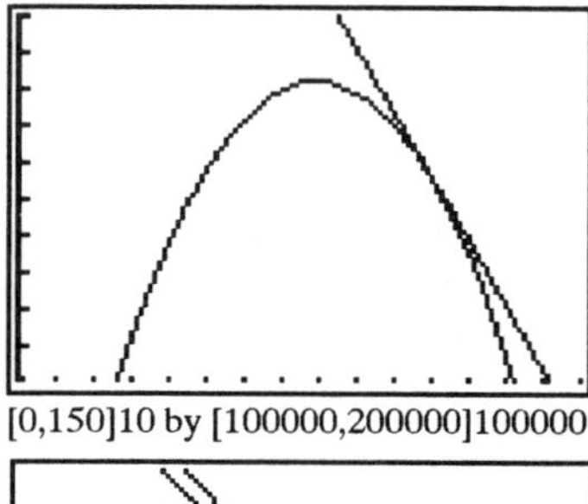
[0,150]10 by [100000,200000]100000

Examining the graph we see there appears to be one point of intersection. This is where the cost function and the revenue function have the same value. We want to find the X coordinate of this point.

Use trace and zoom or change the viewing window variables to [104, 114]0 by [150000, 160000]0 to see that the graphs really do not intersect. This means that the revenue will never equal the cost. There is no break-even point. Since the revenue will always be less than the cost, this would not be a good product to produce based on this analysis.

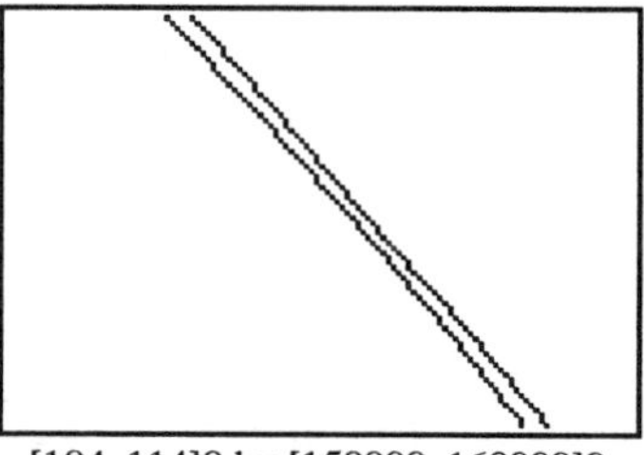
[104, 114]0 by [150000, 160000]0

This can be verified by solving the equation $71000 + 60(4600 - 29p) = (4600 - 29p)p$. This can be done algebraically by hand or using the built-in features on the TI-82 or TI-85 calculator. See Appendix Section B-9 or C-9 of this manual.

Exercise Set 1

1. Evaluate the functions in Problems 13-30 of Textbook Exercise 1-1.

2. Graph the function and use trace and zoom to approximate the domain of Problems 41-50 of Textbook Exercise 1-1.

3. Evaluate the expression in Problem 75 of Textbook Exercise 1-1 for $x = 2.5$, 2.1, 2.01, and 2.001. Examine the results to determine what value $Q(x)$ gets close to as x gets close to 2.

4. Repeat Problem 3 above for $x = 1$, 1.5, 1.9, 1.99, and 1.999. Examine the results to determine what value $Q(x)$ gets close to when x gets close to 2.

5. Let $f(x) = x^2 - 2.3x + 3$. How does the quantity $\frac{f(1.5+h) - f(1.5)}{h}$ change as h assumes values .2, .1, .01, and .001. Fill in the chart and make a conjecture as to what happens to this quantity when h gets closer and closer to zero.

h	.2	.1	.01	.001	→ 0
$\frac{f(1.5+h) - f(1.5)}{h}$					

6. Let $f(x) = 3x^2 - 1.3x + 3$. How does the quantity $\frac{f(.8+h) - f(.8)}{h}$ change as h assumes values .2, .1, .01, and .001. Fill in the chart and make a conjecture as to what happens to this quantity when h gets closer and closer to zero.

h	.2	.1	.01	.001	$\rightarrow$	0
$\frac{f(.8+h) - f(.8)}{h}$						

7. Refer to Problem 90 of Textbook Exercise 1-1. Find the error between the price-demand function $p(x)$ that models the data and the price p for each value of x.

8. Graph the functions in Problems 21-28 of Textbook Exercise 1-2 and explain how each is related to the graph of one of the six basic functions in Figure 1 of Textbook Section 1-2.

9. Graph the lines in Problems 9-22 of Textbook Exercise 1-3.

10. Graph the lines in Problem 23-26 of Textbook Exercise 1-3.

11. Solve Problem 27 of Textbook Exercise 1-3. Conjecture, then check, what the solution to Part (E) would be if the function was $g(x) = 2.2x - 4.2$.

12. Solve Problem 28 of Textbook Exercise 1-3. Conjecture, then check, what the solution to Part (E) would be if the function was $f(x) = -0.8x + 4.2$.

13. Solve Problem 61 of Textbook Exercise 1-3. Conjecture, then check, if the graph of $3x + 2y = 0$ is in this family of lines.

14. Solve Problem 62 of Textbook Exercise 1-3. Conjecture, then check, how the graphs of $2x + 3y = 0$ and $2x - 3y = 0$ are related.

15. Use graphing techniques to explore whether or not the two lines $5.23x - 8.92y = 6.34$ and $5.21x - 8.89y = 6.30$ are parallel.

16. Use graphing techniques to explore whether or not the two lines $5.23x - 8.92y = 6.34$ and $y = .60x + 6.38$ are parallel.

17. Graph $f(x) = x^2$, $f(x) = 3x^2$, and $f(x) = -1.5x^2$ on the same coordinate axes. Compare the graphs and describe the effect of multiplying $f(x) = x^2$ by a constant.

18. Graph $f(x) = x^2$, $f(x) = (x-1)^2$, and $f(x) = (x+2)^2$ on the same coordinate axes. Compare the graphs and describe the effect of adding/subtracting a constant from the x term of $f(x) = x^2$.

19. Refer to Problem 67 of Textbook Exercise 1-3. Suppose the price-demand table was as follows:

Demand (x)	Wholesale Price $p(x)$
0	$250
2,300	180
4,800	115
7,800	80

(A) Plot these points, letting $p(x)$ represent the price at which x number of mowers can be sold (demand).

(B) Draw a line that models the data as closely as possible. (See Appendix Section A-3 & A-16, B-3 & B-16, or C-3 & C-16 of this manual.)

(C) Estimate, to two decimal places, the coefficients of the line and write the equation of the line. (See Appendix Section A-3, B-3, or C-3 of this manual).

(D) Use the line that models the data to predict the wholesale price if the demand is 3,500. (See Appendix Section A-4, B-4, or C-4 of this manual.)

20. Repeat Problem 15 above using the following price-demand table:

Demand (x)	Wholesale Price $p(x)$
0	$300
2,300	190
4,800	110
7,800	50

21. Suppose that in Example 12 above the demand equation is $x = 9000\text{-}30p$ and the cost equation is changed to $C = 90000+30x$.

(A) Express the cost C as a linear function of the price p.
(B) Express revenue R as a quadratic function of price p.
(C) Graph the cost and revenue functions found in parts A and B on the same set of axes, and identify the regions of profit and loss.
(D) Find the break-even points to the nearest dollar.
(E) Find the price that produces the maximum revenue.

22. Repeat Problem 12 above if the demand equation is $x = 5000 - 50p$ and the cost equation is $C = 40000 + 12x$.

23. Solve Problems 29-34 of Textbook Exercise 1-4.

24. Solve Problems 35-38 of Textbook Exercise 1-4.

25. Solve Problems 41-44 of Textbook Exercise 1-4.

26. Do Textbook Chapter 1 Group Activity. Refer to Appendix Section A-16, B-16, or C-16 for instructions on how to use the calculator to find the least squares linear regression line.

NOTES

Chapter 2

Additional Elementary Functions

This chapter contains examples using the calculator to illustrate:

- Determining the domain of a function
- Exploring discontinuities of a function
- Exploring vertical asymptotes of a function
- Exploring horizontal asymptotes of a function
- Exploring properties of graphs
- Approximating solutions of equations in one variable
- Approximating solutions of inequalities in one variable
- Evaluating exponential and logarithmic functions
- Graphing exponential and logarithmic functions
- Solving problems involving growth or decay
- Graphing inverse functions
- Using the change of base formula to evaluate logarithms

The graphing calculator features used are:

- Storing a function
- Evaluating a function
- Graphing a function
- Trace
- Zoom
- Determining the viewing window variables based on the number of pixels on the calculator screen.

EXAMPLE 1 ***TEXTBOOK SECTION 2-1***

Given the function $f(x) = \frac{x^2-4}{x-2}$.

(A) Graph using [-4.7, 4.8]1 by [-1, 5.3]1 on the TI-81, [-4.7, 4.7]1 by [-1, 5.2]1 on the TI-82, or [-6.3, 6.3]1 by [-1, 5.2]1 on the TI-85. Note that by using this range the coordinates for x and y at the bottom of the viewing window are in steps of .1 when you use the arrow keys to find screen locations (not trace) of the calculator. This is because the screen on the TI-81 is 96 pixels across by 64 pixels down; on the TI-82 it is 95 pixels across by 63 pixels down; and on the TI-85 the screen is 127 pixels across by 63 pixels down.

(B) Find the domain.

Solution (A): Enter the function in the function list as (X^2-4)÷(X-2). Set the viewing window variables and graph. Note that there is a hole in the graph at x=2. Use the trace feature of the calculator to see this.

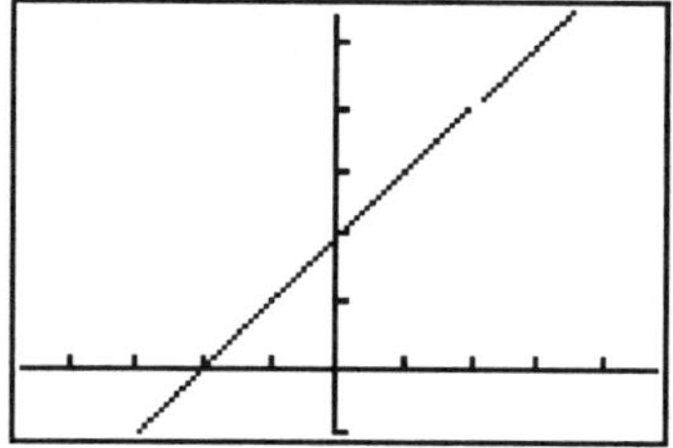

Solution (B): When using the trace feature of the calculator you see that there is no coordinate for y at the bottom of the screen when the x coordinate is 2. This indicates that 2 is not in the domain of the function.

Another way to see this is to return to the home screen, store 2 as X, and evaluate the function. This evaluation causes an error message to occur. This indicates that the function cannot be evaluated at this value of x. Notice that the denominator of $f(x)$ is equal to zero when x=2. Since we cannot divide by zero, x=2 must be ruled out of the domain. Hence the domain of $f(x)$ is $D = \{x \mid x \neq 2\}$.

EXAMPLE 2 ***TEXTBOOK SECTION 2-1***

Consider the function $f(x) = \dfrac{3x^2 + 5x - 8}{x^2 - 1}$.

(A) Graph using [-4.7, 4.8]1 by [-12.8, 12.4]1 on the TI-81, [-4.7, 4.7]1 by [-12.4, 12.4]1 on the TI-82, or [-6.3, 6.3]1 by [-12.4, 12.4]1 on the TI-85. Note that by using this range the coordinates for x at the bottom of the viewing window are in steps of .1 and y is in steps of .4 when you use the trace function of the calculator. This is because the screen on the TI-81 is 96 pixels across by 64 pixels down; on the TI-82 it is 95 pixels across by 63 pixels down; and on the TI-85 the screen is 127 pixels across by 63 pixels down.

(B) Find the domain.

Solution (A): Enter the function in the function list as (3X^2+5X-8)÷(X^2−1). Set the viewing window variables and graph the function. Note that there is unusual behavior when x =-1 and x =1 (a hole in the graph).

Solution (B): Use the trace function of the calculator to see that there is no y coordinate when the x coordinate is 1 or -1. This indicates that -1 and 1 are not in the domain of the function.

Another way to see this is to return to the home screen, store 1 as X, and evaluate the function. An error message occurs indicating that the function cannot be evaluated at this value of x. This can also be seen be examining the denominator of $f(x)$. The denominator is equal to zero when x=1. Similarly for x = -1.

Hence the domain of $f(x)$ is $D = \{x \mid x \neq 1, x \neq -1\}$.

EXAMPLE 3 ***TEXTBOOK SECTION 2-1***

(A) Investigate the behavior of $f(x) = \dfrac{3x^2 + 5x - 8}{x^2 - 1}$ when x is close to 1 and -1.

(B) Write the summary statement symbolically as is done in the solution to Textbook Section 2-1 Example 2.

(C) Write the equations of any vertical asymptotes.

(D) Investigate the graph of f as x increases or decreases without bound.

Solution: We shall conduct our investigation by constructing a table for $f(x)$ for x values near but not equal to the given value and by examining the graph of the function using the trace and zoom features of the calculator.

Solution (A): Store the function in the calculator as (3 X ^ 2 + 5 X - 8) ÷ (X ^ 2 - 1). Store .9 as X and evaluate the expression. The result is 5.632, rounded to the nearest thousandth. Repeat for the other values of x.

x	.900	.990	.999	→ 1 ←	1.001	1.010	1.100
$f(x)$	5.632	5.513	5.501→ ?		?←5.499	5.488	5.381

The table shows that the function is approaching 5.5 as x approaches 1 from both the right and the left. Hence, $\lim_{x \to 1} f(x) = 5.5$. However, as we saw in Example 2 above, when 1 is stored for x and the function is evaluated we get an error message. The denominator has value 0 when x=1 and the calculator is attempting to divide by zero. This indicates that the function is not defined for x=1. Hence, x=1 is not in the domain of the function.

Set the calculator to dotted mode. (See Appendix Section A-14, B-14, or C-14 Example 2 of this manual.) Set the viewing window variables to [-10,10]1 by [-10,10]1 and graph the function that is stored. Use trace and move the cursor along the curve for values of x near 1. Note the y values are near 5.5. Now zoom in on the point of the graph where x is close to 1 or change the viewing window variables to [.90,1.10].01 by [5.3,5.7].1 and trace again. We do not see a hole in the graph displayed on the calculator screen since the x values on the screen do not equal 1 exactly. The graph will show a hole at x = 1 if the viewing window variables are set as in Example 2 above .

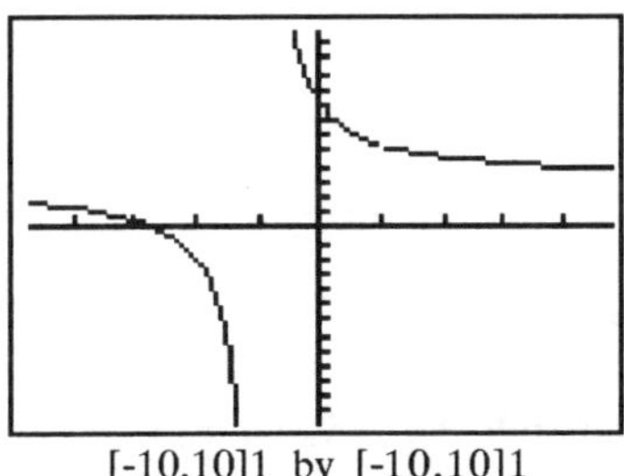

[-10,10]1 by [-10,10]1

Repeat the investigation for values of x near -1. Make sure the function is still stored in your calculator.

x	-1.100	-1.010	-1.001	→ -1 ←	-.999	-.990	-.900
$f(x)$	-47	-497	-4997→ ?		?←5003	503	53

The table shows that the function values are getting greater and greater negatively for values of x close to but a little less than -1. The function values are getting larger and larger positively for values of x close to but a little greater than -1.

Set the viewing window variables to [-10,10]1 by [-10,10]1 and graph the function that is stored. Use trace and move the cursor along the curve for values of x near -1. We see that the y values get very large in the negative direction as x gets close to but is less than -1. Then as the cursor is moved from a little less than -1 to x values a little greater than -1, the y values jump to very large positive numbers. However, as we saw in Example 2 above, when -1 is stored for x and the function is evaluated we get an error message. The denominator has value 0 when x=-1 and the calculator is attempting to divide by zero. This indicates that the function is not defined for x=-1. Hence, x=-1 is not in the domain of the function.

Solution (B): Symbolically we write $f(x) \to -\infty$ as $x \to -1^-$ $f(x) \to \infty$ as $x \to -1^+$.

Solution (C): The line x = -1 is a vertical asymptote.

Solution (D): To investigate the function as x increases without bound let us make a table. Store the function as Y1 in the calculator. Store the value of 10 as X and evaluate the function. (See Appendix Section A-4, B-4, or C-4 of this manual.)

x	10	100	1,000	10,000	→	∞
$f(x)$	3.4545	3.0495	3.0050	3.0005	→	3

The graph of $y = f(x)$ approaches the line y=3 from above as x increases without bound. This can also be seen by examining the graph above or using trace and using the right arrow keys to move the cursor far to the right. Watch the values of the y coordinate as you do this.

We repeat the process to investigate the function as x decreases without bound. We complete the table below.

x	-10	-100	-1,000	-10,000	→	-∞
$f(x)$	2.4444	2.9495	2.9950	2.9995	→	3

The graph of $y = f(x)$ approaches the line y=3 from below as x decreases without bound. This can also be seen by examining the graph above or using trace and using the left arrow keys to move the cursor far to the left. Watch the values of the y coordinate as you do this.

EXAMPLE 3 ***TEXTBOOK SECTION 2-1***

Consider the function $f(x) = \frac{|x-2|}{x-2}$. (A) Investigate the behavior of x near 2.
(B) Does $f(x)$ have a discontinuity at x=2?

Solution (A): We shall conduct our investigation by constructing a table for $f(x)$ for x values near but not equal to 2 and by examining the graph of the function for x near 2 using the trace and zoom features of the calculator.

Store the function in the calculator as ABS (X - 2) ÷ (X - 2) where ABS is the built-in absolute value function in your calculator. Store each value of x in the calculator and evaluate the stored function.

x	1.900	1.990	1.999	→ 2 ←	2.001	2.010	2.100
$f(x)$	-1	-1	-1 → ?	?←	1	1	1

We see that the function values are -1 for values of x less than 2. The function values are 1 for values of x greater than 2.

Set the calculator in dotted mode. Set the viewing window variables to [-10,10]1 by [-10,10]1 and graph. Use trace and move the cursor near x=2. Watch the y values as x moves from a little less than 2 to a little greater than 2. The y value jumps from -1 to 1.

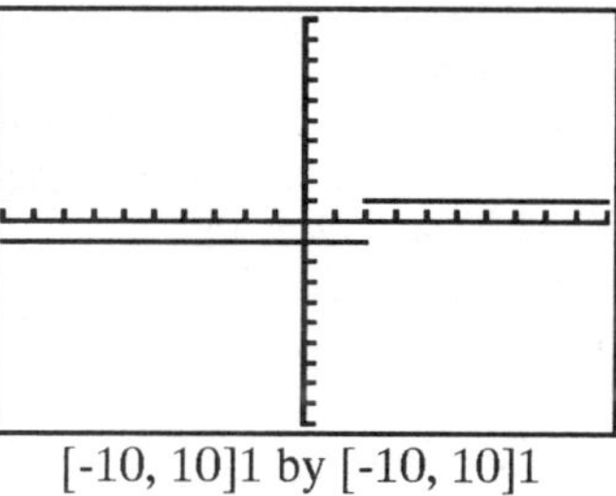
[-10, 10]1 by [-10, 10]1

Solution (B): Substitute 2 for x in the calculator and evaluate the function. We get an error message. This is because the calculator is trying to evaluate $\frac{0}{0}$.
Hence we see that $f(x)$ is discontinuous at x = 2.

EXAMPLE 4 ***TEXTBOOK SECTION 2-2***

Approximate to two decimal places the solution of $3x^2 = \frac{1}{x^3+2} + 4x^4$ for $^-5 \le x \le 5$.

The **solution to an equation in one variable** can be approximated using <u>two different methods regardless of the equation.</u>

<u>METHOD 1</u>
Set the equation equal to zero.
Graph.

(We want to find the x values where the expression has the value zero. These are the x intercepts.)

Find the x intercepts.
The x intercepts are the solutions to the equation.

METHOD 2
Graph the left side of the equation and the right side of the equation as two separate functions.
(We want to find the x values where these two expressions are equal. These values occur at the points of intersection.)
Find the x value of the points of intersection.
These x values are the solutions to the equation.

Solution: METHOD 1

Setting the expression equal to zero we get $3x^2 - \left(\frac{1}{x^3+2} + 4x^4\right) = 0.$

Enter the expression in the calculator as the function: (3X^2)-(1÷(X^3+2)+4X^4). Graph using [-10,10]1 by [-10,10]1. (See Appendix Sections A-6 & A-7, B-6 & B-7, or C-6 & C-7 of this manual.)

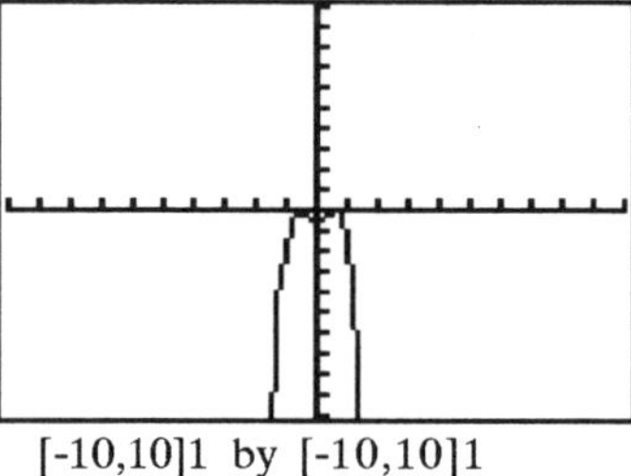

[-10,10]1 by [-10,10]1

We see that the function crosses the x axis near the origin. Set the viewing window variables to [-1,1].1 by [-1,1].1. The graph is shown below at the right. There appear to be three values of x that make this expression equal to 0. Approximate the x intercepts. Use any of the methods explained in Appendix Section A-7, B-7, or C-7 of this manual. For example, in order to use METHOD 1 set the viewing window variables to [-.6,-.5].01 by [-.1,.1].01. Then change to [.4,.5].01 by [-.1,.1].01. Finally change to [.7,.8].01 by [-.1,.1].01.

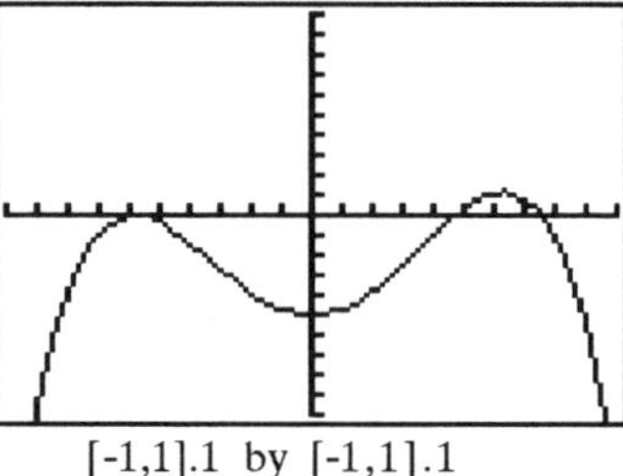

[-1,1].1 by [-1,1].1

Use trace and zoom in each case to approximate the four solutions: $x \approx -.60$, $x \approx -.56$, $x \approx .48$, and $x \approx .75$.

Solution: METHOD 2

Store the left and right expressions as separate functions in the calculator as: 3X^2 and 1÷(X^3+2)+4X^4. Set the viewing window variables to [-10,10]1 by [-10,10]1. Graph. We need to zoom in to approximate the coordinates of the points of intersection. Change the viewing window variables or use the zoom-in features of the calculator. (See Appendix Section A-7, B-7, or C-7 of this manual.) Let us simply change the viewing window variables to get a better look at the points of intersection. We see that the functions cross in the first and second quadrants. Set the viewing window variables to [-1,1]1 by [-1,3]1.

We need to zoom in again. Change the viewing window variables to [-.6,-.5].01 by [.9,1.2].1, then [.4,.5].01 by [0,1].1 and finally [.7,.8].01 by [1,2].1 or use other zoom-in features of the calculator. Use trace to find the x values of the points of intersection. They are $x \approx -.60$, $x \approx -.56$, $x \approx .48$, and $x \approx .75$.

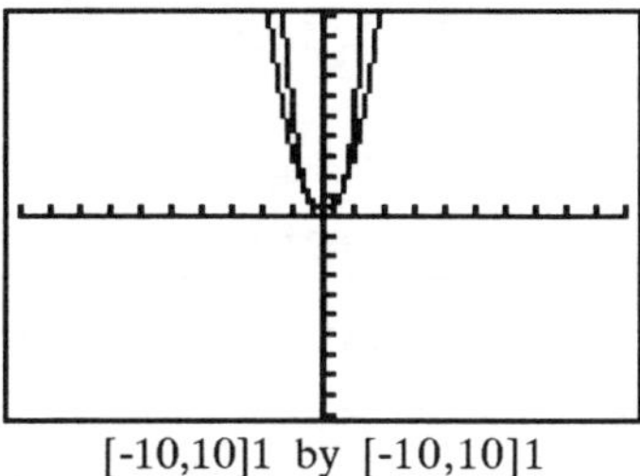

[-10,10]1 by [-10,10]1

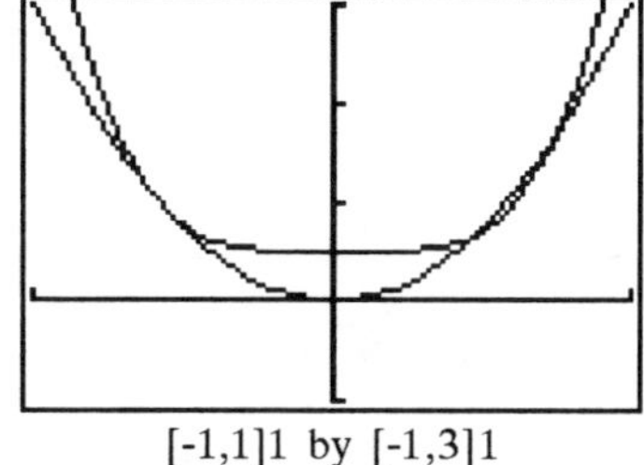

[-1,1]1 by [-1,3]1

EXAMPLE 5 ***TEXTBOOK SECTION 2-2***

Approximate to two decimal places the solution of $3x^2 < \dfrac{1}{x^3+2} + 4x^4$ for $-5 \le x \le 5$.

The **solution to an inequality in one variable** can be approximated using one of two methods. Both of the methods start with graphing the function as in Example 4 above. Then the interval on x must be determined.

METHOD 1

Change the inequality sign to an equation sign.
Set the equation equal to zero.
Graph. (We want to find the set of x values which makes the inequality a true statement. We first must find the values of x where the expression is equal to zero. These values on x are where the graph crosses the x axis. These are the x intercepts.)
Find the x intercepts.
Determine the interval on x that makes the inequality true.
(This can be done by inspecting the graph.)

METHOD 2

Graph the left side of the inequality and the right side of the inequality as two separate functions. (We want to find the set of x values which makes the inequality a true statement. First we want to find the x values where these two expressions are equal. These values occur at the points of intersection.)
Find the x value of the points of intersection.
Determine the interval on x that makes the inequality true.
(This can be done by inspecting the graph.)

Solution: METHOD 1

The expression was set equal to zero and the x intercepts were found in METHOD 1 of Example 4 above. They are -.60, -.56, .48, and .75. Inspecting the graph we see it is below the x axis for $x < -.60$, $-.56 < x < .48$, and $.75 < x$. Hence the solution to the inequality written in interval notation is $(-\infty,-.60)\cup(-.56,.48)\cup(.75,\infty)$.

Solution: METHOD 2

The points of intersection were found in METHOD 2 of Example 4 above. The x values are -.60, -.56, .48, and .75. Inspecting the graph we see that the graph of the left side of the inequality is below the graph of the right side of the inequality for $x < -.60$, $-.56 < x < .48$, and $.75 < x$. Hence the solution to the inequality written in interval notation is $(-\infty,-.60)\cup(-.56,.48)\cup(.75,\infty)$.

EXAMPLE 6 *TEXTBOOK SECTION 2-2*

Graph $f(x) = 1.2^x$, $g(x) = 2.5^x$, $h(x) = -4^x$, and $k(x) = 4^x$ on the same set of coordinate axes. Describe the differences between the graphs.

Solution: Store the four functions in the calculator as 1.2^X, 2.5^X, -4^X, and 4^X. Note that -4^x is not the same function as $(-4)^x$. In fact, -4 cannot be the base of an exponential function.

Graph first using [-10,10]1 by [-10,10]1. It is hard to see the difference between the graphs. Change the viewing window variables to [-3,3]1 by [-5,5]1 to get the graph shown at the right.

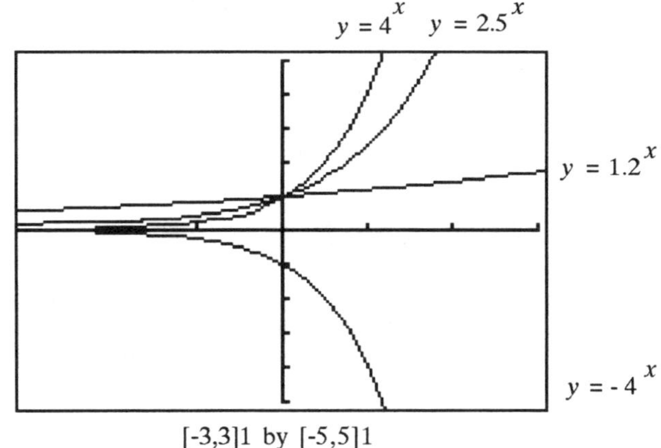

[-3,3]1 by [-5,5]1

The graph $f(x) = 1.2^x$ is lower than $g(x) = 2.5^x$ when $x > 0$, equal $\log(x) = 2.5^x$ when $x=0$ and higher than $g(x) = 2.5^x$ when $x < 0$. A similar comparison can be made between $g(x) = 2.5^x$ and $k(x) = 4^x$. The graph of $h(x)$ is reflected about the x axis from $k(x)$.

EXAMPLE 7 *TEXTBOOK SECTION 2-2*

A group of 400 parents, relatives, and friends are waiting anxiously at an airport for a student charter flight to return after a four-week trip to Europe. It is stormy and the plane is late. A particular parent thought he had heard that the plane's radio had gone out and related this news to some friends, who in turn passed it on to others, and so on. Sociologists have studied rumor propagation and have found that a rumor tends to spread according to: $N(t) = \dfrac{400}{1+399e^{-0.4t}}$ where t is measured in minutes.

(A) Graph this function.
(B) How many people will have heard the rumor after 5 minutes?
(C) Approximate the amount of time for 50% of the people to have heard the rumor.
(D) Approximate the amount of time for 75% of the people to have heard the rumor.
(E) Approximate the amount of time it will take for everyone to hear the rumor.

Solution: (A) Change t to X and store the function into the calculator as: 400 ÷ (1 + 399 **e^** (⁻ .4 X)). (Recall e^ is the inverse **ln** function.)

We need to determine the viewing window variables. Evaluate the function for some values of X. Round numbers up since $N(x)$ represents the number of people having heard the rumor. For example, $N(10) = 48.146$ means that 49 people heard the rumor.

x	$N(x)$
0	1
10	49
20	353
30	400

From this table we see a good starting viewing window is [0,50]5 by [0,500]100.

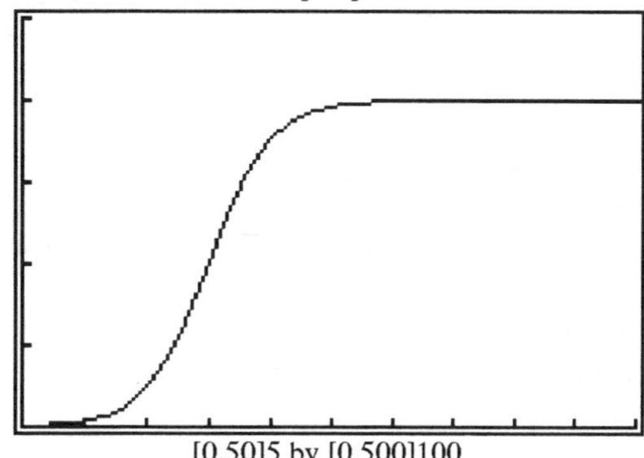

[0,50]5 by [0,500]100

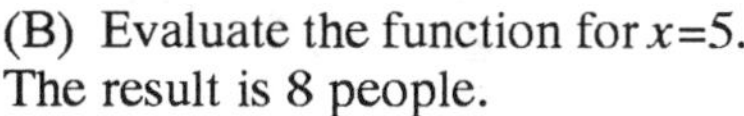

(B) Evaluate the function for x=5. The result is 8 people.

(C) Use trace and move the cursor to a position on the curve where the y value is close to 200 since 200 is 50% of 400. Zoom in on this value until accuracy is to one decimal place and round up. The value of x is approximately 15. It takes approximately 15 minutes for one-half of the people to have heard the rumor.

(D) Use trace and move the cursor to a position on the curve where the y value is close to 300 since 300 is 75% of 400. Zoom in on this value until accuracy is to one decimal place and round up. The value of x is approximately 18. It takes approximately 18 minutes for three-fourths of the people to have heard the rumor.

(E) Change the viewing window variables back to [0,50]5 by [0,500]100. Now use trace and watch the y variable as x increases. We see that as x gets larger and larger the y value gets closer and closer to 400. This makes sense since all people will hear a rumor eventually. We also see that 400 is reached when $x \approx 32$ which means that everyone will have heard the rumor in about 32 minutes.

EXAMPLE 8 ***TEXTBOOK SECTION 2-3***

(A) Graph $y = 3^x$, its inverse, and $y = x$ on the same set of coordinate axes. (B) Calculate the y coordinate for $y = 3^x$ when $x = 2$. Calculate the y coordinate for $y = \log_3 x$ when $x = 9$. Mark the points on the respective graphs. Discuss the relationship between these two pairs of coordinates.

Solution (A): First we need to write the inverse function of $y = 3^x$. We want to graph $y = \log_3 x$. The calculator can only calculate or graph functions having base 10 or base *e*. However, there is a formula that allows us to change from one base, say *b*, to another base, say *c*. This formula is: $\log_b x = \frac{\log_c x}{\log_c b}$. In our situation we have:

$\log_3 x = \frac{\log_{10} x}{\log_{10} 3}$. Hence we will store the three functions separately in the calculator as 3^X, X, and ([LOG] X) ÷ ([LOG] 3). Graph using [-6,6]1 by [-4,4]1. Note the y axis limits are 2/3 those of the x axis. This will make the scale marks about the same distance apart on both axes. The graph is shown at the right below.

Solution (B): For $y = 3^x$, $y = 9$ when $x = 2$. So the first coordinate pair is (2, 9).

For $\log_3 x$, $y = 2$ when $x = 9$. The second coordinate pair is (9, 2).

The coordinates are interchanged between the function and its inverse. The points are shown on the graph above. They are equidistant from the line $y = x$.

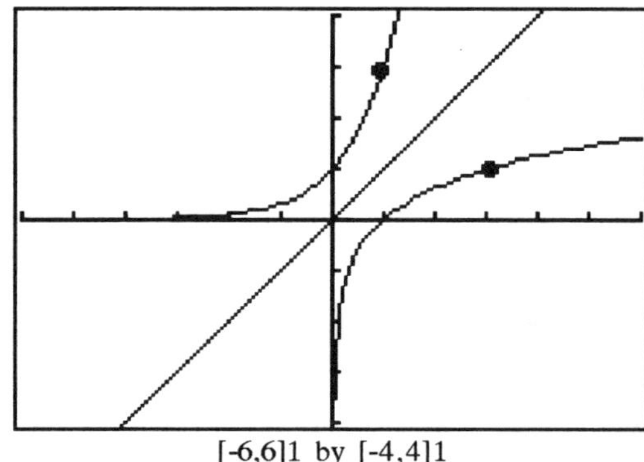

[-6,6]1 by [-4,4]1

Exercise Set 2

1. Solve Problems 15-20 of Textbook Exercise 2-1.
2. Solve Problems 21-24 of Textbook Exercise 2-1 by graphing the functions to answer the questions.
3. Solve Problems 25-28 of Textbook Exercise 2-1.
4. Solve Problems 29-34 of Textbook Exercise 2-1.
5. Solve Problems 39-40 of Textbook Exercise 2-1. To answer Part (D), graph the function and use trace to investigate what happens to $f(x)$ as x increases.
6. Solve Problems 1-12 of Textbook Exercise 2-2.
7. Solve Problems 27-36 of Textbook Exercise 2-2.
8. Solve Problems 37-40 of Textbook Exercise 2-2.

9. Solve Problems 47-50 of Textbook Exercise 2-2 using graphing methods.

10. Solve Problems 51-54 of Textbook Exercise 2-2.

11. Solve Problems 65-66 of Textbook Exercise 2-2.

12. Solve Problems 67-72 of Textbook Exercise 2-2.

13. Approximate the solution to two decimal places of $5x^2+3 = \frac{2}{x^3-1}$.

14. Approximate the solution to two decimal places of $3x^2 + 4x = \frac{1}{x^2+1}$.

15. Approximate the solution to two decimal places of $3x^2 + 4x > \frac{1}{x^2+1}$.

    Use viewing window variables [-3,3]1 by [-3,3]1.

16. Approximate the solution to two decimal places of $5x^2+3 < \frac{2}{x^3-1}$.

17. Graph $f(x) = 2.1^{x-1}$, $g(x) = 2.1^{x-2}$, $h(x) = 2.1^{x+1}$, and $j(x) = 2.1^x$ on the same graphing screen. Describe the differences between the graphs.

18. Graph $f(x) = 2.1^x$, $g(x) = 2.1^{3x}$, and $h(x) = 2.1^{-2x}$ on the same graphing screen. Describe the differences between the graphs.

19. Evaluate to five decimal places $\log_{1.3}368.93$.

20. Evaluate to five decimal places $\log_{12.3}368.93$.

21. Graph $y = 4^x$, its inverse, and $y = x$ on the same set of coordinate axes.

22. Graph $y = 1.5^x$, its inverse, and $y = x$ on the same set of coordinate axes.

23. Solve Problems 95-96 of Textbook Exercise 2-3.

24. Do Textbook Chapter 2 Group Activity. Use the calculator where appropriate.

NOTES

Chapter 3

Mathematics of Finance

This chapter contains examples using the calculator to illustrate:

- Calculation of simple interest, compound interest, future value of an ordinary annuity, and present value of an ordinary annuity
- Graphing of the future value with simple interest and compound interest
- Approximation of the interest rate in future value or amortization
- Evaluation of an algebraic expression to find future value and amortization schedules

The graphing calculator features used are:

- Calculation
- Replay key
- Ans key
- Storing an algebraic expression
- Evaluating an algebraic expression
- Graphing
- Trace and zoom

EXAMPLE 1 ***TEXTBOOK SECTION 3-1***

$38,597 is invested in an account earning simple interest of 3.57%.

(A) How much interest will be earned if the money is left in the account for one year? two years? five years?

(B) Graph the future value as a function of time.

> The replay key allows rapid calculation of expressions that are similar. This key will replay the last executed expression so you can change a quantity and recalculate. (TI-81: Press the Up Arrow. TI-82 and TI-85: Press [2nd] [ENTRY].)

Solution (A): Enter 38597×.0357×1 in the calculator and evaluate. The result is $1377.91. Use the replay key and change the 1 to a 2. Evaluate again. The result is $2755.83. Use the replay key again and change the 2 to a 5. Evaluate again. The result is $6889.56.

Solution (B): The future value of this loan after t years is A = 38597(1+.0357t). Change the t to an X and store the function as Y1 in the calculator. An appropriate setting for the viewing window variables would be [0, 10]1 by [0, 15000]1000. The graph is shown at the right.

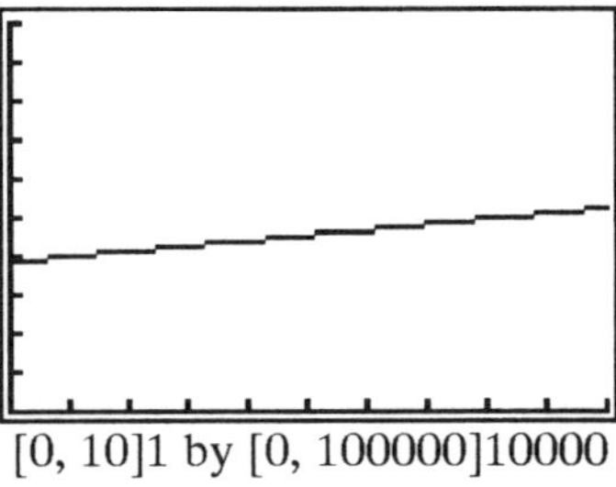
[0, 10]1 by [0, 100000]10000

Note that this graph implies interest is accumulating continuously when in fact it is calculated at the end of each time period.

EXAMPLE 2 ***TEXTBOOK SECTION 3-2***

$4,570 is invested at 7.8% compounded monthly. (A) Make a chart showing the amount in the account at the end of each period for eight months. (B) How much interest is earned each month? (C) How much interest is earned overall? (D) How does the interest in the first month relate to that in each of the succeeding months? (E) Repeat Parts (A) through (D) under the conditions that the interest is withdrawn from the account at the end of each month. (F) Graph the future value as a function of time for each of the situations in this example. Why is the graph for the situation in Part (E) a horizontal line?

Solution (A) & (B): Store the expression $P(1+i)^n$ in the calculator as a function in the calculator as P(1+I)^N. Store P×I as a separate function in the calculator. Store 4570 as P, .078÷12 as I, and 1 as N.

Calculate the second expression. The result is $29.71.
Calculate the first expression. The result is $4599.71.
Store this result as P.

Calculate the second expression again. The result is $29.90
Calculate the first expression again. The result is $4629.60.
Store this result as P.

Repeat this procedure until the chart has been filled in.

Period Number	Principal at the start of the month	Amount at the end of the month	Amount of interest earned each period
1	4570	4599.71	29.71
2	4599.71	4629.60	29.90
3	4629.60	4659.70	30.09
4	4659.70	4689.98	30.29
5	4689.98	4720.47	30.48
6	4720.47	4751.15	30.68
7	4751.15	4782.03	30.88
8	4782.03	4813.12	31.08

Sum = $243.12

Solution (C): The sum of the interest amounts should be equal to the value of the expression $P(1+i)^n - P$ for P=4750, I=.078/12, and N=8. Make certain you still have the expression $P(1+i)^n$ stored in the calculator as P(1+I)^N. Store 4570 as P, .078÷12 as I, and 8 as N. Evaluate the expression. Storing these values in the calculator and evaluating the stored expression yields $4813.12 Now subtract P. The result is $243.12. This is the same amount as the sum of the "Amount of interest earned each period" column in the table on the preceding page.

Solution (D): The interest on each succeeding month is the interest from the previous month plus the interest on this amount. For example: The interest for month 2 is $29.71 + 29.71 \times \frac{.078}{12}$. This demonstrates that the interest also earns interest each succeeding month.

Solution (E): The chart is now:

Period Number	Principal at the start of the month	Amount at the end of the month	Amount of interest earned each period
1	4570	4599.71	29.71
2	4570	4599.71	29.71
3	4570	4599.71	29.71
4	4570	4599.71	29.71
5	4570	4599.71	29.71
6	4570	4599.71	29.71
7	4570	4599.71	29.71
8	4570	4599.71	29.71
			Sum = $29.71x8 = $237.68

When the interest earned during a month is withdrawn at the end of the month, the principal remains the same each month and hence, the interest earned in each month is the same.

Solution (F): The expression for the first situation is $4570\left(1+\frac{.078}{12}\right)^x$ for integer values of x so that the interest is calculated and added to the principal at the beginning of the next month. The graph is shown to the right.

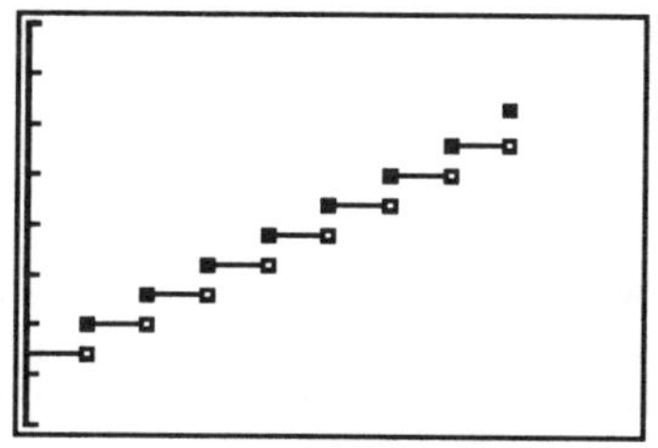

The graph can be displayed on the calculator by entering (4570 (1 + .078 ÷ 12) ^ iPart X) as Y1 and using [0, 9.5]1 by [4500, 4900]0 on the TI-81, [0, 9.5]1 by [4500, 4900]0 on the TI-82 and [0, 12.6]1 by [4500, 4900]0 on the TI 85. iPart is found on the MATH menu. The open and closed dots are not displayed on the calculator.

In the second situation, the amount in the account is 4570 for the entire month until the end of the month when the interest is calculated. The interest is added to the account at the beginning of the next month but then immediately withdrawn. This graph can be displayed using the same viewing window variables. Store 4570 (1 + .078 ÷ 12) (X = iPart X) as Y1 and 4570 ÷ (X ≠ iPart X) as Y2 where the = and ≠ signs are from the TEST menu and iPart is from the MATH menu. The open and closed dots are not displayed on the calculator.

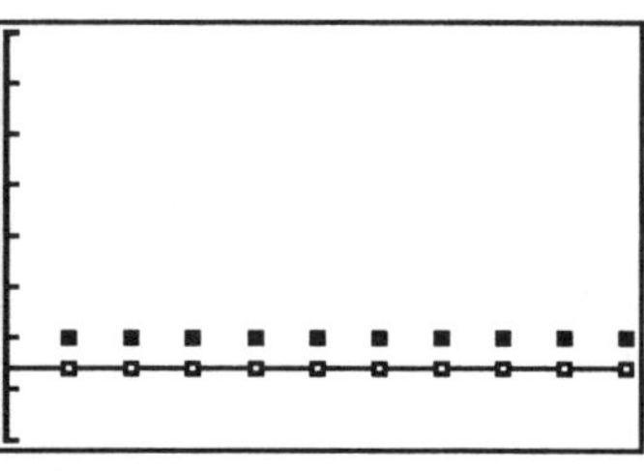

[0, 10]1 by [4500, 4900]0

However, we usually graph these as continuously increasing functions. To do this enter 4570 (1 + .078 ÷ 12) ^ X as Y1 in the calculator for the first situation. Enter 4570 ÷ (X < 1) as Y2 and (4570 (1 + .078 ÷ 12)) ÷ (X > 1) as Y3 and for the second situation. The inequality signs are found on the TEST menu. Set the viewing window dimensions at [0, 10]1 by [4500, 4900]0. Graph. The open and closed dots are not displayed on the calculators.

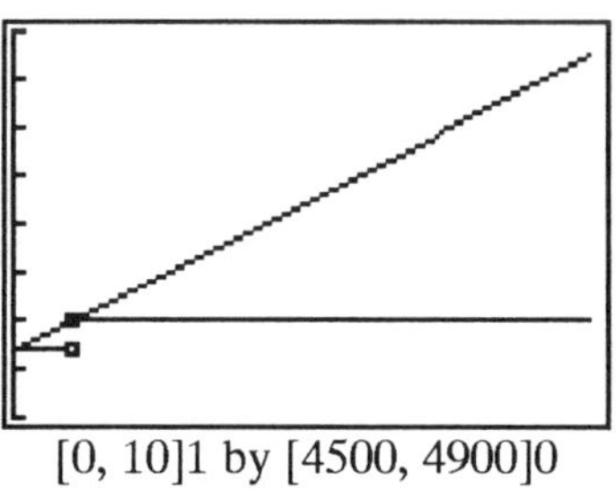

[0, 10]1 by [4500, 4900]0

The graphs are shown to the right. Note that the future value starts at $4570 since both situations start with the same principal. The graph in the second situation is a horizontal line since the future value for each month is $4599.71. The principal does not change each month, as in the first situation, since the interest is withdrawn from the account at the end of each month.

EXAMPLE 3 ***TEXTBOOK SECTION 3-2***

Graph the growth of an investment of $5000 at 10% compounded semiannually and the growth of an investment of $4500 at 12% compounded daily. (A) Which investment would be better if you plan to leave the money invested for a period of 10 years? (B) How much time will have elapsed before the investments yield the same amount?

Solution (A): We must first store the appropriate expressions in the functions list in the calculator. Enter 5000 (1 + .10 ÷ 2) ^ (2 x X) as Y1 for the first investment. Enter 4500 (1 + .12 ÷ 365) ^ (365 x X) as Y2 for the second investment. Set the viewing window variables at [0, 10]1 by [5000, 25000]500. The graphs are shown to the right. The investment of $4500 at 12% is the better investment if money is left in the account for 10 years.

Solution (B): Using trace and zoom, or other calculator methods, we find that the point of intersection occurs at x=4.7 years. Since the first investment calculates interest each 6-months, the investments would be said to be equal after 5 years.

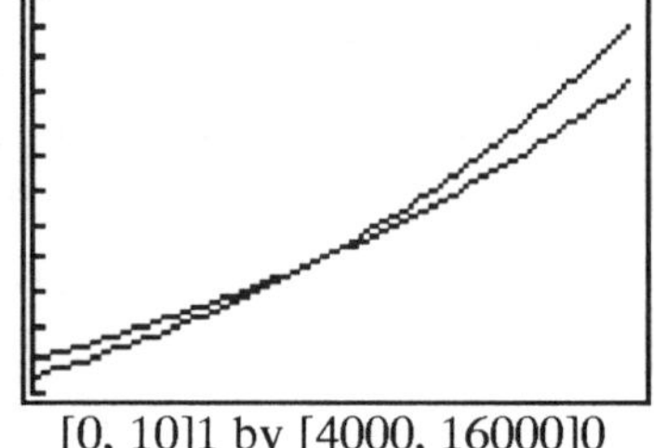

[0, 10]1 by [4000, 16000]0

EXAMPLE 4 ***TEXTBOOK SECTION 3-3***

\$150 is deposited each month in an account earning 5.6% interest compounded monthly. Deposits were started five years ago. How much interest was earned during the third year?

Solution: We need to find the amount that was in the account at the end of the second year. Then we need to find the amount that was in the account at the end of the third year. The amount of interest earned during the third year is then given by

$$\left(\begin{array}{c}\text{the amount in the account}\\ \text{at the end of the 3rd year}\end{array}\right) - \left(\begin{array}{c}\text{the amount in the account}\\ \text{at the end of the 2nd year}\end{array}\right) - \left(\begin{array}{c}\text{the amount deposited}\\ \text{the third year}\end{array}\right)$$

To find the amount in the account at the end of the second year calculate

$$S = 150\,\frac{\left(1 + \frac{.056}{12}\right)^{24} - 1}{\frac{.056}{12}}.$$ Enter this into the calculator as: 150 ((1 + .056 ÷ 12) ^ 24 - 1) ÷ (.056 ÷ 12). The result is \$3799.98.

To find the amount in the account at the end of the third year calculate

$$S = 150\,\frac{\left(1 + \frac{.056}{12}\right)^{36} - 1}{\frac{.056}{12}}.$$ The result is 5865.25. The amount deposited during the third year is \$150 x 12 = \$1800.

Hence the total interest earned in the third year is \$5865.25 – 37.99.98 – 1800 = \$265.27.

EXAMPLE 5 ***TEXTBOOK SECTION 3-3***

A sinking fund has been accumulating money for five years. \$250 is deposited each month. Currently there is \$17353.15 in the account. What is the interest rate?

Solution: The formula is $S = R\,\frac{(1+i)^n - 1}{i}$. We are looking for $r = 12i$. We wish to find i from this equation. However, this problem cannot be done algebraically. i can be approximated graphically. Substituting the numbers into the equation for S, R, and n we have $17353.15 = 250\,\frac{(1+i)^{60} - 1}{i}$. Store the left side of this equation as a function as 17353.15 in the calculator. Store the right side of this equation as a second function as: 250 ((1 + X) ^ 60 - 1) ÷ X where X represents i.

We now need to determine values for the viewing window variables. We know i will be between 0 and 1 since it is the monthly interest rate. So the viewing window variables for the horizontal axis will be [0, 1]1.

Since the first function stored is 17353.15 we must scale the y axis to accommodate this large number. Hence graph using [0, 1].1 by [0, 20000]1000. However, when this viewing window is used, nothing of the second function is seen on the graph. Hence we must adjust the variables. We suspect that perhaps the first screen dimensions are too big. Changing the viewing window to [0, .1].01 by [0, 20000]1000 gives the graph on the left below (see next page).

Using trace we see that the point of intersection is near (.005, 17500). Using zoom or changing the range to [0, .05].01 by [17000, 18000]1000, gives the graph on the right. (Note: It may take the calculator a little time to graph.)

Continue using trace and zoom or changing the range until you find $x \approx 0.00483$. This is the value of i. The rate we are seeking is $12i = .058$ or 5.8%.

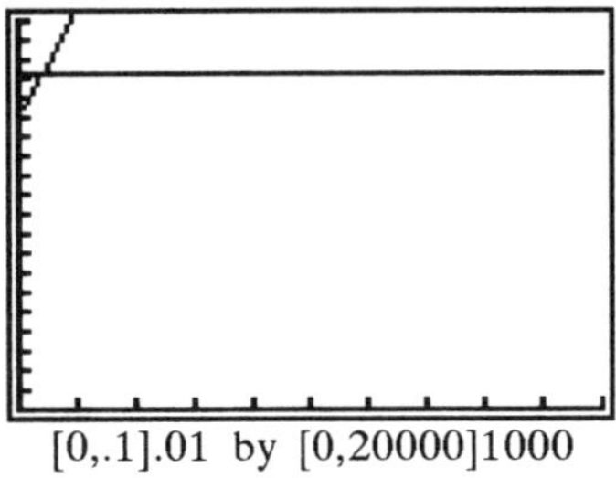

[0,.1].01 by [0,20000]1000

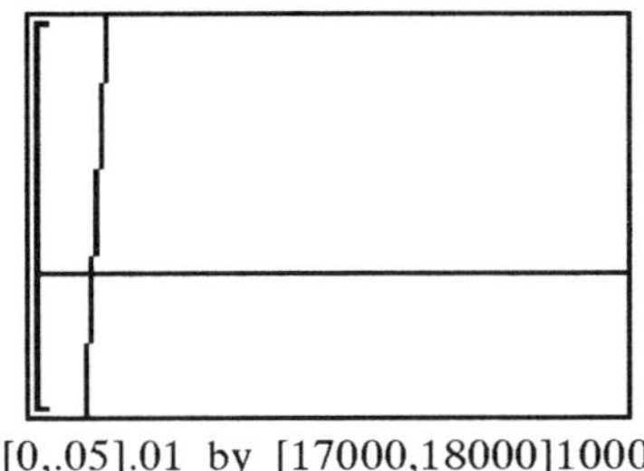

[0,.05].01 by [17000,18000]1000

EXAMPLE 6 ***TEXTBOOK SECTION 3-4***

Suppose you have purchased a car for $11,500. The rate of financing is 9.8% compounded monthly. (A) What are your payments if you finance for one year? (B) Make an amortization schedule. (C) How much interest do you pay if you did not pay off the loan early? (D) Revise the table to reflect the situation if you made $1500 payments instead those calculated in Part (A). (E) How much interest is paid in the situation of Part (D)? (F) Find the payment number after which no more than one-half of the original unpaid balance is due.

Solution (A): Since we are dealing with money in this problem, it may be helpful to set the number of decimal places in the calculator to two. (See Appendix Section A-17, B-17, and C-17 of this manual.)

The formula for finding the payment is $P = 11500 \dfrac{\frac{.098}{12}}{1 - \left(1 + \frac{.098}{12}\right)^{-12}}$. Enter this into the calculator as: 11500 ((.098 ÷ 12) ÷ (1 - (1 + .098 ÷ 12) ^- 12))

The result of the calculation is 1009.96. The payments are $1009.96.

Solution (B): The amortization schedule can be found as follows:

Store the functions $P \times I$, $A - P \times I$, and $P - (A - P \times I)$ as three separate functions in the calculator. Store 11500 as P, .098÷12 as I, and 1009.96 as A.

Evaluate the three stored functions. The results are the interest, unpaid balance reduction, and unpaid balance, respectively. The results of the first calculation are 93.92, 916.04, and 10583.96. Store Ans as *P*. Repeat the instructions in this paragraph, recording the results as you go until the following table is complete.

Payment Number	Payment	Interest on Unpaid Balance	Unpaid Balance Reduction	New Unpaid Balance
1	1009.96	93.92	916.04	10583.96
2	1009.96	86.44	923.52	9660.43
3	1009.96	78.89	931.07	8729.37
4	1009.96	71.29	938.67	7790.70
5	1009.96	63.62	946.34	6844.36
6	1009.96	55.90	954.06	5890.30
7	1009.96	48.10	961.86	4928.44
8	1009.96	40.25	969.71	3958.73
9	1009.96	32.33	977.63	2981.10
10	1009.96	24.35	985.61	1995.48
11	1009.96	16.30	993.66	1001.82
12	1009.96	8.18	1001.78	.04

Note that the unpaid balance is not 0. This is because of rounding that takes place during the calculations. This amount probably would be added to your last payment.

Solution (C): The amount of interest is $1009.96×12 - 11500 = $619.56.

Solution (D): The payment schedule is:

Payment Number	Payment	Interest on Unpaid Balance	Unpaid Balance Reduction	New Unpaid Balance
1	1500.00	11500.00x.098/12= 93.92	1500-93.92= 1406.08	11500.00-1406.08 = 10093.92
2	1500.00	82.43	1417.57	8676.35
3	1500.00	70.86	1429.14	7247.21
4	1500.00	59.19	1440.81	5806.40
5	1500.00	47.42	1452.58	4353.82
6	1500.00	35.36	1464.44	2889.38
7	1500.00	23.60	1476.40	1412.98
8	1500.00	11.54	1488.46	-75.48
9				
10				
11				
12				

We notice that the last value in the New Unpaid Balance column is -75.48. This means that the last payment need only be 1500.00-75.48 or $1424.52.

Solution (E): The amount of interest is (1500.00×7+1424.52) - 11500 = 424.52. This illustrates the value in making payments in excess of those required if your contract will allow you to do so.

Solution (F): Half of the original unpaid balance of $11500 is $5750. We wish to

find x so that $5750 = 1009.96 \dfrac{1 - \left(1+\frac{.098}{12}\right)^{-(12-x)}}{\left(\frac{.098}{12}\right)}$

We see from the table in Part (A) that this occurs between the sixth and seventh payment. Enter 5750 as Y1 in the calculator. Enter 1009.96 ((1 - (1 + .098 ÷ 12) ^ - (12 - X)) ÷ (.098 ÷ 12)) as Y2 in the calculator. Set the viewing window variables to [0, 8]1 by [4000, 7000]1000. Graph. The result is shown to the right.

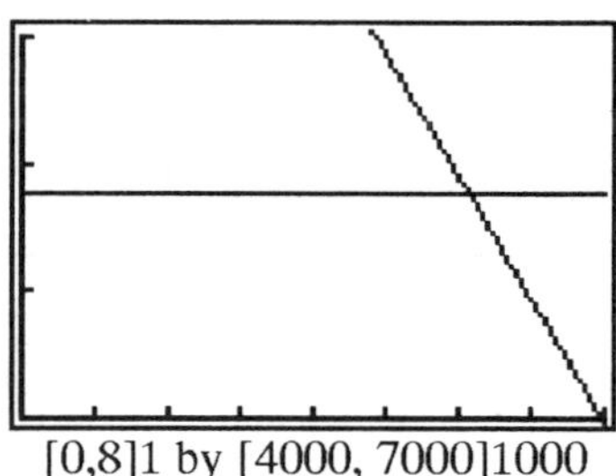
[0,8]1 by [4000, 7000]1000

Use trace and zoom, or other calculator methods, to find the x coordinate of the point of intersection. Since this is not an integer, we round up. Hence the number of payments needed is seven. The unpaid balance after the seventh payment will be a little less than one-half the original unpaid balance.

Exercise Set 3

1. Solve Problems 19-30 of Textbook Exercise 3-1.

2. Solve Problems 31-40 of Textbook Exercise 3-1.

3. $43,865 is invested in an account earning simple interest of 4.96%. How much interest will be earned if the money is left in the account for one year? two years? five years?

4. $8,392 is invested in an account earning simple interest of 7.83%. How much interest will be earned if the money is left in the account for one year? two years? five years?

5. Solve Problems 17-20 of Textbook Exercise 3-2.

6. Solve Problems 23-42 of Textbook Exercise 3-2.

7. Solve Problems 43-44 of Textbook Exercise 3-2.

8. Solve Problems 45-64 of Textbook Exercise 3-2.

9. $6,925 is invested at 6.5% compounded quarterly. (A) Make a chart showing the amount in the account at the end of each quarter for 18 months. Include a column on the chart showing how much interest is earned each quarter. (B) Make a chart showing the amount of the investment at the end of each quarter for 18 months if the interest is withdrawn after it is calculated. (C) Make a graph comparing these two situations.

10. $200 is invested at 5.8% compounded monthly. (A) Make a chart showing the amount in the account at the end of each month for 9 months. Include a column on the chart showing how much interest is earned each quarter. (B) Make a chart showing the amount of the investment at the end of each quarter for 18 months if the interest is withdrawn after it is calculated. (C) Make a graph comparing these two situations.

11. Solve Problems 1-10 of Textbook Exercise 3-3.

12. Solve Problems 11-20 of Textbook Exercise 3-3.

13. Solve Problems 21-26 of Textbook Exercise 3-3.

14. Solve Problems 27-32 of Textbook Exercise 3-3.

15. $50 is deposited each month in an account earning 4.8% interest compounded monthly. Deposits were started five years ago. How much interest was earned during the third year?

16. $375 is deposited each month in an account earning 7.2% interest compounded daily. Deposits were started five years ago. How much interest was earned during the third year?

17. A sinking fund has been accumulating money for five years. $200 is deposited each month. Currently there is $13532 in the account. What is the interest rate?

18. Solve Problems 11-22 of Textbook Exercise 3-4.

19. Solve Problems 23-32 of Textbook Exercise 3-4.

20. Solve Problems 33-34 of Textbook Exercise 3-4.

21. Solve Problems 35-38 of Textbook Exercise 3-4.

22. You bought a new car for $12,500. Your payments are $743.25 for 18 months. Find the interest rate?

23. A stereo system has been purchased for $3500. The rate of financing is 9.8% compounded monthly. What are your payments if you finance for one year? Make an amortization schedule. How much interest do you pay?

24. Repeat Example 5 above if you finance at 8.5% for 18 months. Do you pay more interest under this plan?

25. Do Textbook Chapter 3 Group Activity. Use the calculator where appropriate.

NOTES

Chapter 4

Systems of Linear Equations; Matrices

This chapter contains examples using the calculator to illustrate:

- Solving of systems of equations in two variables using graphing
- Solving of systems of equations using Gauss-Jordan reduction
- Solving systems of equations using the inverse coefficient matrix
- Solving problems using matrix operations

The graphing calculator features used are:

- Graphing
- Trace and zoom
- Matrix row operations
- Matrix operations
- Inverse matrices

EXAMPLE 1 ***TEXTBOOK SECTION 4-1***

Given the system of equations $\begin{aligned} 11x + 2y &= 1 \\ 2x + 5y &= -24 \end{aligned}$.

(A) Solve to two decimal places using graphing techniques.

(B) Conjecture, then check, the effect on the solution of changing the 11 in the first equation to 20.

(C) Conjecture, then check, the effect on the solution of changing the 1 in the first equation of the original system to 6.

Solution (A): Each of these equations must be solved algebraically for y in order to enter it into the calculator.

The system becomes:

$$y = \frac{1-11x}{2}$$

$$y = \frac{-24-2x}{5}$$

Store in the calculator as two separate functions:

(1 - 11 X) ÷ 2

(-24 - 2 X) ÷ 5

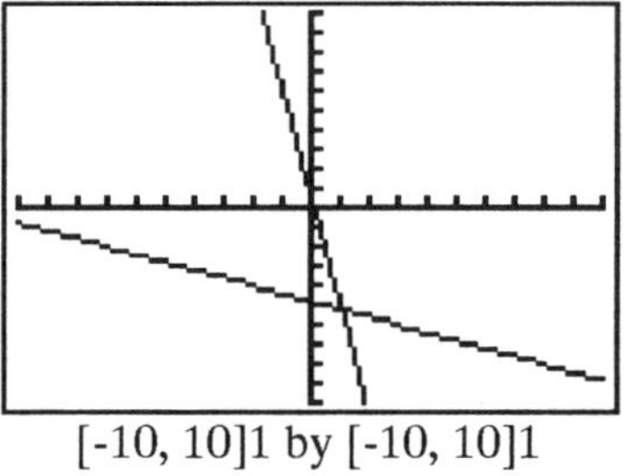

[-10, 10]1 by [-10, 10]1

Graph the lines using [-10,10]1 by [-10,10]1. Use trace and zoom, or other calculator techniques, to find the coordinates of the point of intersection. The solution is $x = 1.04$, $y = -5.22$.

Solution (B): Change the 11 in the first equation to 20. Store this in the calculator as Y3 as (1 - 20 X) ÷ 2. Changing the 11 in the first equation to 20 changes the slope of the line from $-\frac{11}{2} = -5.5$ to $-\frac{20}{2} = -10$. This means the line has steeper slope but the same y intercept.

Hence the x coordinate of the point of intersection will move to the left and the y coordinate will move upward since the slope of the second line is negative. The coordinates of the point of intersection are (.55, -5.02).

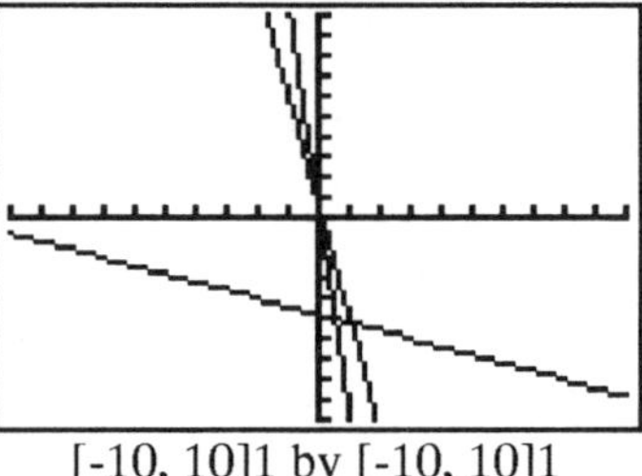

[-10, 10]1 by [-10, 10]1

(All three lines are graphed to the right.)

Solution (C): Change the 1 to 6 in the original first equation. Store this in the calculator as Y3 as (6 - 11 X) ÷ 2. Changing the 1 to 6 in the original first equation changes the y intercept from $\frac{1}{2}$ to $\frac{6}{2} = 3$. This shifts the first line upward (or to the right). Hence the point of intersection will move to the right and downward. The x coordinate will be greater and the y coordinate will be less than the coordinates of the original intersection point. The solution is $x = 1.53$, $y = -5.41$.

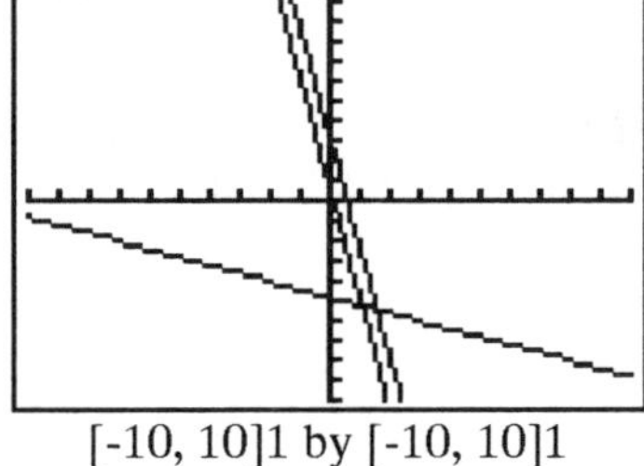

[-10, 10]1 by [-10, 10]1

EXAMPLE 2 ***TEXTBOOK SECTION 4-1***

Solve
$$\begin{aligned} 9x - 3y &= 24 \\ 11x + 2y &= 1 \\ 2x + 5y &= -24 \end{aligned}$$
using graphing techniques.

Solution: Each of these equations must be solved algebraically for y in order to enter it into the calculator.

The system becomes:	Store in the calculator as three separate functions:
$y = \dfrac{24-9x}{-3}$	(24 - 9 X) ÷ -3
$y = \dfrac{1-11x}{2}$	(1 - 11 X) ÷ 2
$y = \dfrac{-24-2x}{5}$	(-24 - 2 X) ÷ 5

Graph these three lines using [-10, 10]1 by [-10, 10]1. The lines appear to intersect in a single point. Use zoom or change the viewing window variables to [.55, 1.55]1 by [-5.8, -4.6]1 to see that the three lines do not intersect in a single point (see graphs next page). Consequently there is no one ordered pair that satisfies all three equations at the same time. Hence there is no solution to this system of equations.

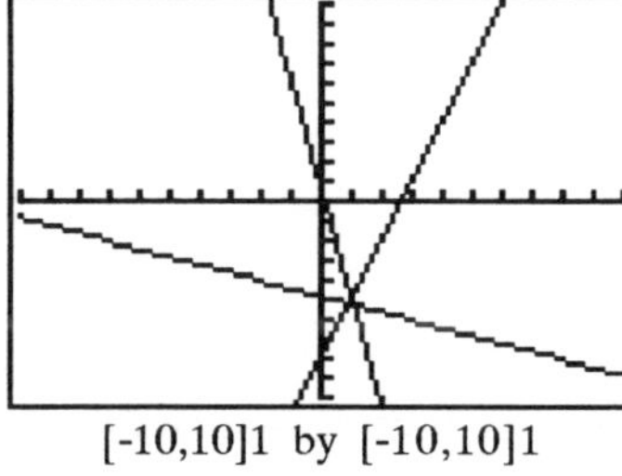
[-10,10]1 by [-10,10]1

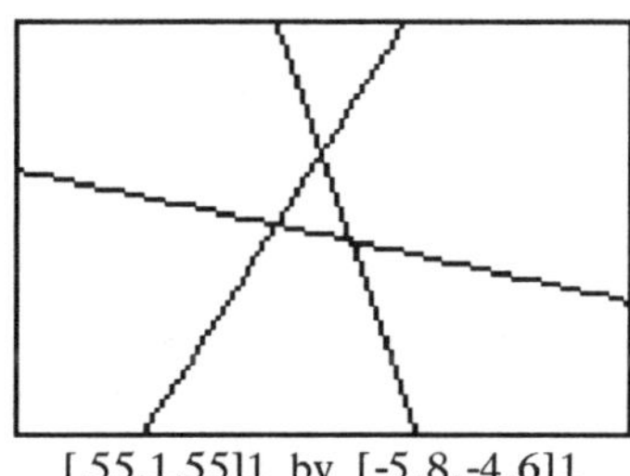
[.55,1.55]1 by [-5.8,-4.6]1

EXAMPLE 3 ***TEXTBOOK SECTION 4-3***

Given the system of equations:

$$\begin{aligned} 7x - 2y + z &= 3 \\ 3x + y - z &= 6 \\ x - 3y + 2z &= 0 \end{aligned}$$

(A) Solve to two decimal places using Gauss-Jordan Elimination.

(B) Conjecture, then check, how the solution for x would be changed if the 3 in the first equation was changed to a 0.

(C) Conjecture, then check, what the solution to the system would be if both the 3 in the first equation and the 6 in the second equation are changed to 0.

Solution (A): Enter the augmented coefficient matrix into the calculator. The particular keystrokes for your calculator are described in Appendix Section A-12, B-12, or C-12 Example 4 of this manual. Use the built-in features of your calculator to perform row operations to get the reduced form.

$$\left[\begin{array}{ccc|c} 7 & -2 & 1 & 3 \\ 3 & 1 & -1 & 6 \\ 1 & -3 & 2 & 0 \end{array}\right] \sim \left[\begin{array}{ccc|c} 1 & 0 & 0 & -1 \\ 0 & 1 & 0 & -19 \\ 0 & 0 & 1 & -28 \end{array}\right].$$

Hence the solution to this system is $x=-1$, $y=-19$, and $z=-28$.

Solution (B): If the 3 is changed to a 0 the value for x will be less than -1. In fact, the solution for x is - 2.

Solution (C): The solution would be $x=0$, $y=0$, and $z=0$ since there would be no values added or subtracted in the constant column.

EXAMPLE 4 — *TEXTBOOK SECTION 4-4*

Solve to two decimal places using inverse matrix methods:

$$\begin{aligned} 7x - 2y + z &= 3 \\ 3x + y - z &= 6 \\ x - 3y + 2z &= 0 \end{aligned}$$

Solution: Enter the coefficient matrix into the calculator. The particular keystrokes for your calculator are described in Appendix Section A-12, B-12, or C-12 Example 2 of this manual. Use the built-in features of your calculator to get the inverse matrix shown below.

$$A^{-1} = \begin{bmatrix} .3333 & -.3333 & -.3333 \\ 2.3333 & -4.3333 & -3.3333 \\ 3.3333 & -6.3333 & -4.3333 \end{bmatrix}$$

Find the matrix product of A^{-1} and B:

$$\begin{bmatrix} .3333 & -.3333 & -.3333 \\ 2.3333 & -4.3333 & -3.3333 \\ 3.3333 & -6.3333 & -4.3333 \end{bmatrix} \begin{bmatrix} 3 \\ 6 \\ 0 \end{bmatrix} = \begin{bmatrix} -1 \\ -19 \\ -28 \end{bmatrix}.$$ Hence $x=-1$, $y=-19$, and $z=-28$.

EXAMPLE 5 ***TEXTBOOK SECTION 4-4 & 4-5***

Given the matrices below find: (A) AB (B) BA (C) $A + A^{-1}$ (D) $2.18A$

$$A = \begin{bmatrix} .28 & .31 & .45 \\ .06 & -.02 & -.40 \\ .09 & -.42 & -.93 \end{bmatrix} \qquad B = \begin{bmatrix} .47 \\ -6.38 \\ .81 \end{bmatrix}$$

Use two decimal place accuracy in all answers.

Solution: Store matrices A and B in the calculator. (See Appendix Section A-12, B-12, or C-12 of this manual.)

(A) Find the product AB. The result is $\begin{bmatrix} -1.48 \\ -.17 \\ 1.97 \end{bmatrix}$.

(B) Find the product BA. An error message appears on the calculator screen. This is because the dimensions are not correct for multiplication. You cannot multiply a 3×1 matrix by a 3×3 matrix.

(C) Find the sum $A + A^{-1}$. The result is $\begin{bmatrix} 3.51 & -1.84 & 2.94 \\ -.37 & 6.49 & -3.41 \\ .60 & -3.57 & -.41 \end{bmatrix}$.

(D) Find 2.18A . The result is $\begin{bmatrix} .61 & .68 & .98 \\ .13 & -.04 & -.87 \\ .20 & -.92 & -2.03 \end{bmatrix}$.

EXAMPLE 6 ***TEXTBOOK SECTION 4-4***

The time requirements for the manufacture of slalom and trick water skis is given by :

	Fabricating department	Finishing department
Trick ski	6.0 hr	1.5 hr
Slalom ski	4.0 hr	1.0 hr

$= A$

Now suppose the company has two manufacturing plants, X and Y, in different parts of the country and that their hourly rates for each department are given in the following matrix:

$$\begin{array}{r} \\ \text{Fabricating department} \\ \text{Finishing department} \end{array} \begin{array}{c} \text{Plant X} \quad \text{Plant Y} \\ \begin{bmatrix} \$8/\text{hr} & \$7/\text{hr} \\ \$6/\text{hr} & \$4/\text{hr} \end{bmatrix} \end{array} = B$$

(A) Find the total labor costs for each ski at each factory.

(B) Conjecture, then check, how do the total labor costs change if the hourly rates at Plant X in the fabricating department and the finishing department increased to \$8.50/hr and \$6.35/hr respectively?

***Solution** (A)* Enter the two matrices into the calculator. (See Appendix Section A-12, B-12, or C-12 of this manual.) Find AB. The result is $\begin{bmatrix} \$57 & \$48 \\ \$38 & \$32 \end{bmatrix}$.

***Solution** (B):* A good conjecture would be that the total labor costs would increase since both the hourly rates increased. Change the matrix elements in matrix B as indicated above and recalculate. The result is now is $\begin{bmatrix} \$60.53 & \$48 \\ \$40.35 & \$32 \end{bmatrix}$. The labor costs at Plant X increase \$3.53 per ski for the trick ski and \$2.35 per ski for the slalom ski.

Exercise Set 4

1. Solve Problems 1-34 of Textbook Exercise 4-1 using graphing techniques.

2. Solve Problems 35-46 of Textbook Exercise 4-1 using graphing techniques.

3. Solve $\begin{aligned} 7x - 5y &= 14 \\ 11x + 2y &= 1 \\ 2x + 5y &= -10 \end{aligned}$ using graphing techniques.

4. Solve $\begin{aligned} 7x - 5y &= 14 \\ 11x + 2y &= 1 \\ 2x + 3.5y &= -7.5 \end{aligned}$ using graphing techniques.

5. Conjecture, then check, how the solution to Exercise 3 above will change if the constant in the first equation is 16 instead of 14.

6. Conjecture, then check, how the solution to Exercise 4 above will change if the coefficient on x in the first equation is 6 instead of 7.

7. Solve Problems 11-16 of Textbook Exercise 4-2 using the calculator. See Appendix Section A-12, B-12, or C-12 of this manual on how to operate your calculator.

8. Solve Problems 17-22 of Textbook Exercise 4-2.

9. Solve Problems 23-44 of Textbook Exercise 4-2.

10. Solve Problems 45-50 of Textbook Exercise 4-2.

11. Solve to two decimal places using Gauss-Jordan elimination:

$$\begin{aligned} 6x - 3y + 2z &= 3 \\ 3x + 2y - 3z &= 6.1 \\ x - 3y + z &= 5.2 \end{aligned}$$

12. Solve to two decimal places using Gauss-Jordan elimination:

$$\begin{aligned} 5x + 3y + 2z &= 4.5 \\ 3.1x - 2y + z &= 6.2 \\ 3x - 3.7y + 2.2z &= 9 \end{aligned}$$

13. Solve Problems 1-18 of Textbook Exercisc 4-4.

14. Solve Problems 19-44 of Textbook Exercise 4-4.

15. Solve Problems 1-14 of Textbook Exercise 4-5.

16. Solve Problems 15-24 of Textbook Exercise 4-5.

17. Solve Problems 25-38 of Textbook Exercise 4-5.

18. Solve Problems 43-46 of Textbook Exercise 4-5.

19. Solve Problems 13-20 of Textbook Exercise 4-6.

20. Solve Problems 27-32 of Textbook Exercise 4-6.

21. Solve to two decimal places using inverse matrix methods:

$$\begin{aligned} 5x + 3y + 2z &= 4.5 \\ 3.1x - 2y + z &= 6.2 \\ 3x - 3.7y + 2.2z &= 9 \end{aligned}$$

22. Solve to two decimal places using inverse matrix methods:

$$\begin{aligned} 6x - 3y + 2z &= 3 \\ 3x + 2y - 3z &= 6.1 \\ x - 3y + z &= 5.2 \end{aligned}$$

23. Repeat all parts of Example 5 in this chapter of the manual if the hourly rate for the fabricating department at Plant X begins at \$7.50/hr instead of \$8/hr. Again, describe what happens when Plant X costs rise to \$8.50/hr and \$6.35/hr for fabricating and finishing, respectively.

24. Suppose a third type of ski was to be manufactured needing 8 hours in the fabricating department and 2 hours in the finishing department. Repeat Example 5 in this manual above under these conditions.

25. Do Textbook Chapter 4 Group Activity. Use the calculator where appropriate.

Chapter 5

Linear Inequalities and Linear Programming

This chapter contains examples using the calculator to illustrate:

- Solving of systems of inequalities in two variables using graphing
- Finding corner points graphically
- Solving a linear programming problem in two variables using graphing
- Solving linear programming problems using matrix methods

The graphing calculator features used are:

- Graphing
- Trace and zoom
- Matrix row operations

EXAMPLE 1 ***TEXTBOOK SECTION 5-1***

Consider the system of linear inequalities listed below: (A) Solve the system graphically. (B) Find the corner points. (C) Conjecture, then check, the effect on the solution if the 10 in the first inequality was changed to 5. (D) Conjecture, then check, the effect on the solution if the 2.3 in the original system of inequalities was changed to a 1.

$$\begin{aligned} 2.3x + y &\geq 10 \\ 3x + y &\leq 14 \\ 1.8x + 4.8y &\leq 48.3 \\ x &\geq 0 \\ y &\geq 0 \end{aligned}$$

Solution (A): Step 1 Change the inequality signs to equal signs and graph the lines. This results in graphing the lines:

$$\begin{aligned} y &= 10 - 2.3x \\ y &= 14 - 3x \\ y &= \frac{48.3 - 1.8x}{4.8} \\ x &= 0 \\ y &= 0 \end{aligned}$$

Store the first three of these expressions in the calculator as separate functions: 10 - 2.3 X, 14 - 3 X, and (48.3 - 1.8 X) ÷ 4.8 . Graph using [0,15]1 by [0,15]1. We know the graph is in the first quadrant because of the last two inequalities.

Step 2 Change the expressions stored as Y1, Y2, and Y3 into statements with the appropriate inequality sign. (See Appendix Section A-5, B-5, or C-5 of this manual.) To do this use the INS function on the calculator and get the inequality signs from the TEST menu. The expressions are now stored in Y1, Y2, and Y3 as: Y ≥ 10 - 2.3 X, Y ≤ 14 - 3 X, and Y ≤ (48.3 - 1.8 X) ÷ 4.8.

Step 3 Choose a point not on any of the lines. We will use (1, 1). Enter 1 for X and 1 for Y in the calculator. Evaluate the inequalities by recalling each of the expressions stored in the calculator. If the result is 1 the statement is true and the point is in the solution region of that inequality. If the result in the calculator is 0, then the statement is not true and the point is not in the solution region. In this case we want the side of the line on the opposite side from (1, 1).

We see that the first inequality is false for $x=1$ and $y=1$. The second and third inequalities are true. So we want the region above the first line and below the second and third lines. This is shown in the figure to the right.

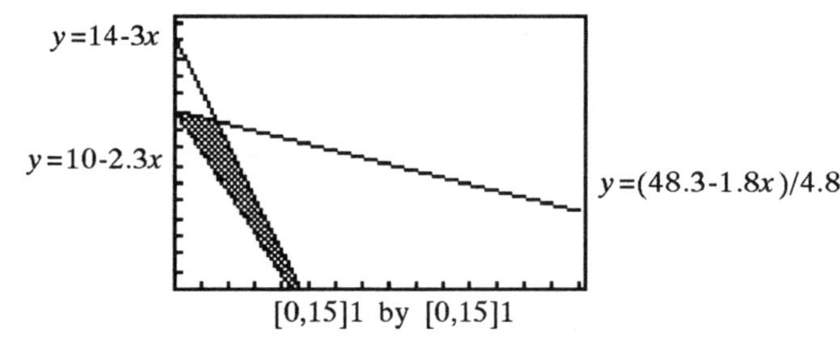

The solution of this system of linear inequalities is the set of ordered pairs corresponding to points in the shaded region and including the line segments that bound it. The shading cannot be done on all graphing calculators. It is shown here to indicate the solution region.

Solution (B): Change the expressions stored as Y1, Y2, and Y3 back to those in Step 1 above by deleting the Y and the inequality signs. Graph the lines. The corner points are found by using zoom and trace, or other calculator methods. The two corner points on the y axis are (0, 10) and (0, 10.06). The two corner points on the x axis are (4.35,0) and (4.67,0). The other corner point is (1.50,9.50).

Solution (C): If the first inequality was $2.3x + y \geq 5$, the y intercept of the boundary line would be 5 instead of 10. This would shift the line downward. Since the slope of the line would be the same, the x intercept would shift to the left.

This situation is shown in the graph to the right. The corner points are now (0.00, 5.00), (0.00, 10.06), (2.17, 0.00), (4.67, 0.00) and (1.50, 9.50).

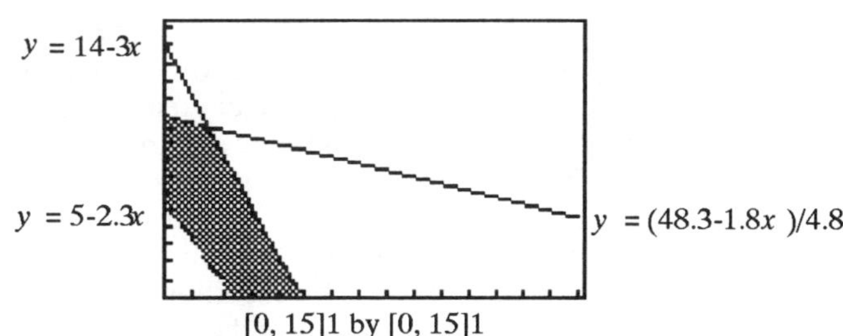

Solution (D): If the first inequality was $x + y \geq 10$, the slope of this boundary line is not as steep as that of the original boundary line. The y intercept remains the same. Hence the x intercept moves to the right. In fact, the x intercept is now 10. Since the x intercept of $y=14\text{-}3x$ is 4.67, the corner point is now the intersection of the two lines $y=14\text{-}3x$ and $y=x+10$. The corner points of the solution region are now (0.00, 5.00), (0.00, 10.06), (2.00, 8.00), and (1.50, 9.50).

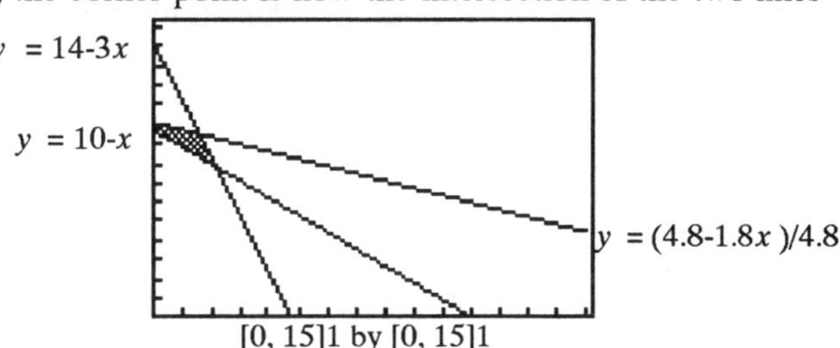

EXAMPLE 2 ***TEXTBOOK SECTION 5-2***

Maximize and minimize the function $z = 1.2x + y$ over the region graphed in Example 1 above.

Solution: We wish to see what the objective function line looks like. Graph the line $1.2x + y = 15$. (The value of 15 was chosen because the largest y intercept was 10.06.) Enter this in the calculator as 15-1.2X. Also graph 14-1.2X, 13-1.2X, 7-1.2X and 3-1.2X on the graph. You may not be able to get all of the lines graphed on the screen at the same time. (See Appendix Sections A-6, B-6, or C-6 of this manual.) Sketch a graph on paper and add the parallel lines.

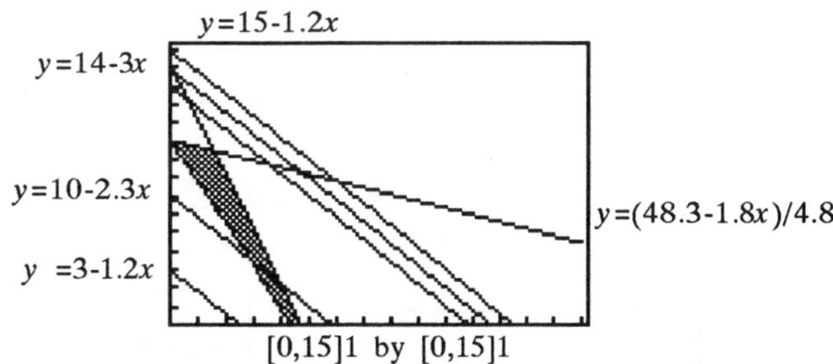

We see that as we decrease the constant value in the objective function line the first corner point encountered is (1.50, 9.50). This will be the point that gives the maximum value to z. Similarly, we see that the smallest value for z will occur at the corner point (4.35, 0). Evaluating z at these points results in: A maximum value of z occurs at (1.5, 9.5). The maximum value of z is 11.3. A minimum value of z occurs at (4.35, 0). The minimum value of z is 5.22.

EXAMPLE 3 **TEXTBOOK SECTION 5-4**

A manufacturer of lightweight mountain tents makes a standard model and an expedition model for national distribution. Each standard tent requires 1 labor-hour from the cutting department and 3 labor-hours from the assembly department. Each expedition tent requires 2 labor-hours from the cutting department and 4 labor-hours from the assembly department. The maximum labor-hours available per day in the cutting department and the assembly department are 32 and 84, respectively. If the company makes a profit of $50 on each standard tent and $80 on each expedition tent, how many tents of each type should be manufactured each day to maximize the total daily profit (assuming all tents can be sold)?

What changes occur if the amount of labor-hours increase .5 for all categories (except the maximum hours available per day)?

Solution: The mathematical model for this problem is:

$$\begin{aligned} \text{Maximize} \quad & P = 50x + 80y \\ \text{Subject to} \quad & x + 2y \le 32 \\ & 3x + 4y \le 84 \\ & x, y \ge 0 \end{aligned}$$

The initial system is:

$$\begin{aligned} x + 2y + s_1 \qquad\qquad &= 32 \\ 3x + 4y \qquad + s_2 \qquad &= 84 \\ -50x - 80y \qquad\qquad + P &= 0 \end{aligned}$$

So the initial tableau is:

	x_1	x_2	S_1	S_2	P	
S1	1	2	1	0	0	32
S2	3	4	0	1	0	84
P	-50	-80	0	0	1	0

Use matrix row operations to perform the pivot operations. (See Appendix Section A-12, B-12 or C-12 of this manual.) The final tableau is:

	x_1	x_2	S_1	S_2	P	
S1	0	1	1.5	-.5	0	6
S2	1	0	-2	1	0	20
P	0	0	20	10	1	1480

So the maximum profit of $1480 is achieved if 20 standard tents and 6 expedition tents are manufactured and sold. Both slack variables have a value of zero. To see this, substitute the values for x and y into the slack variable equations. This means that all the hours available in the two departments are used.

Exercise Set 5

1. Solve Problems 1-12 of Textbook Exercise 5-1.

2. Solve Problems 17-20 of Textbook Exercise 5-1.

3. Solve Problems 25-34 of Textbook Exercise 5-1.

4. Solve Problems 35-44 of Textbook Exercise 5-1.

5. Solve the following system of linear inequalities graphically, and find the corner points:

$$\begin{aligned} 4.5x + 2.1y &\ge 15 \\ 3x + 7.2y &\ge 18.4 \\ 1.8x + 4.8y &\le 48.3 \\ x &\ge 0 \\ y &\ge 0 \end{aligned}$$

6. Solve the following system of linear inequalities graphically, and find the corner points:

$$\begin{aligned} 5.5x + 3.1y &\ge 13.5 \\ 2x + 6.2y &\le 19.2 \\ 1.8x + 4.8y &\le 48.3 \\ x &\ge 0 \\ y &\ge 0 \end{aligned}$$

7. Solve Problems 9-26 of Textbook Exercise 5-2

8. Solve Problems 9-12 of Textbook Exercise 5-4 graphically and by the simplex method.

9. Solve Problems 13-18 of Textbook Exercise 5-4 graphically and by the simplex method.

10. Solve Problems 19-28 of Textbook Exercise 5-4.

11. Solve Problems 31-32 of Textbook Exercise 5-4 graphically and by the simplex method.

12. Solve Problems 33-34 of Textbook Exercise 5-4 graphically and by the simplex method.

13. Maximize and minimize the function $z = 8.2x + 2.1y$ over the region graphed in Problem 5 above.

14. Maximize and minimize the function $z = 3.1x - y$ over the region graphed in Problem 6 above.

15. Repeat the first part of Example 3 above if the profit on standard tents is $90 and on expedition tents $70.

16. Repeat Example 3 if a new type of tent is introduced requiring 3 labor-hours from the cutting department and 4.5 labor-hours from the assembly department. The profit on each new tent is $60. What is the effect of having 38 hours and 90 hours available per day in the cutting department and the assembly department, respectively?

17. Solve Problems 5-20 of Textbook Exercise 5-5 graphically and by applying the simplex method to the dual problem.

18. Solve Problems 21-24 of Textbook Exercise 5-5.

19. Solve Problems 1-12 of Textbook Exercise 5-6 graphically and using the big M method.

20. Solve Problems 13-32 of Textbook Exercise 5-6 using the calculator to perform appropriate row operations.

21. Do Textbook Chapter 5 Group Activity. Use the calculator where appropriate.

Chapter 6

Probability

This chapter contains examples using the calculator to illustrate:

- Finding permutations and combinations
- Calculating probabilities and expected values
- Generating random numbers
- Calculating conditional probabilities
- Applying Bayes' Formula
- Finding expected values

The graphing calculator features used are:

- Replay key
- Calculation
- Permutations and combinations
- Random number generator

EXAMPLE 1 ***TEXTBOOK SECTION 6-2***

A group of 12 people are meeting.

(A) In how many ways can 3 be chosen for one committee and 5 be chosen for a second committee. No person can serve on both committees.

(B) How would this result be affected if there were 5 people chosen for the first committee and 3 for the second committee?

(C) How would the result be affected if there were only 11 people from which to choose?

(D) Suppose the 12 people are numbered from 1 to 12. Use the random number generator in the calculator with a seed number of 6 to pick 7 of these 12 randomly.

Solution (A): This is a problem involving the fundamental theorem of counting and combinations. Use the built-in features of your calculator to find $C_{12,3}C_{9,5}$. The result is 27720 ways.

Solution (B): The calculation would be $C_{12,5}C_{7,3}$ if 5 people were chosen for the first committee and 3 people were chosen for the second committee. The result is the same, 27720 ways.

Solution (C): Use the replay key to change the last entry 12 to 11 and 7 to 6. (On the TI-81 press the up arrow. On the TI-82 and TI-85 press [2nd] [ENTRY].) Recalculate. The result is 9240. By reducing the number of people you start with by one, the number of ways is about one-third as many.

Solution (D): Store the seed number in the storage location RAND (see Appendix Section A-11, B-11, or C-11 of this manual). [IPart] (12 x [Rand]) will generate integers x such that $0 \le x \le 11$. Since we want integers $1 \le x \le 12$, we add 1 to this expression. So we enter [IPart] (12 x [Rand]) +1 in the calculator to get the numbers:

6	2	5	10	2	7	6	9	10
8	10	9	9	1	1	12	12	5
11	7	5	8	2	11	10	1	5

Since we cannot have repeats, the people picked are numbered
6 2 5 10 7 9 8.

EXAMPLE 2 — *TEXTBOOK SECTION 6-3*

(A) Use the random number generator to get a sample of 40 tosses of a nickel and a dime.
(B) How do the results compare to what you expected from an equally likely sample space (see Textbook Section 6-3 Example 4).

Solution (A): Choose a seed number (say 3). Store this in the storage location RAND (see Appendix Section A-11, B-11, or C-11 of this manual). Suppose 0 represents no heads, 1 represents a head on the nickel and a tail on the dime, 2 represents a tail on the nickel and a head on the dime, and 3 represents two heads. We wish to generate random numbers x, $0 \le x \le 3$. Enter [IPart] (4 x [Rand]) in the calculator (see Example 1(D) above). The resulting numbers are:

0	2	2	1	0	1	0	3	1	3
1	3	1	0	0	1	3	0	1	1
2	1	2	1	3	0	0	1	1	2
0	3	2	3	2	2	3	2	3	0

Counting we see the results are: 10 0's, 12 1's, 9 2's, and 9 3's.

Solution (B): The expected results were 10 in each category.

EXAMPLE 3 — *TEXTBOOK SECTION 6-5*

In a study to determine employee voting patterns in a recent strike election, employees were selected at random and the following tabulation was made:

		SALARY CLASSIFICATION		
		Hourly(H)	Salary(S)	Salary+Bonus(B)
TO	Yes(Y)	394	175	20
STRIKE	No(N)	142	111	123

(A) Convert this table to a probability table.
(B) What is the probability of an employee voting to strike (Y) given that the person is paid hourly (H)?
(C) Are the events H and Y independent?

Round all answers to three decimal places.

Solution: (A) Store this data as a 2×3 matrix A. Then calculate the sum of the elements. This can be done by recalling the elements using the features of the calculator, or simply adding the six numbers in the calculation mode. The result is 965. Store this answer as T using Ans→T. Then calculate $\frac{1}{T}[A]$ and store as matrix B. This is the probability matrix.

$$B = \frac{1}{T}\begin{bmatrix} 394 & 175 & 20 \\ 142 & 111 & 123 \end{bmatrix} = \begin{bmatrix} .408 & .181 & .021 \\ .147 & .115 & .127 \end{bmatrix}$$

(B) We wish to calculate $P(Y|H) = \frac{B(1,1)}{B(1,1)+B(2,1)}$. Do this by recalling the elements of matrix B. Reentering the values of matrix B will cause round-off errors. In the calculation mode of the calculator enter **[B](1,1)** ÷ (**[B](1,1)** + **[B](2,1)**) where [B](1,1) represents the way to recall the element in the first row and first column of matrix B. It is shown in bold here to indicate that a single number is stored in the calculator in each location. The result is .735 .

(C) We wish to see if P(Y|H) = P(H). P(Y|H) was calculated in Part (B). Calculate P(H) by calculating (**[A](1,1)** + **[A](2,1)**) ÷ T. The result is .555 . Hence the events H and Y are not independent since .555 ≠ .735 .

EXAMPLE 4 — *TEXTBOOK SECTION 6-6*

A new inexpensive skin test is devised for detecting tuberculosis. To evaluate the test before it is put into use, a medical researcher randomly selects 1,000 people. Using precise but more expensive methods already available, it is found that 8% of the 1,000 people tested have tuberculosis. Now each of the 1,000 subjects is given the new skin test and the following results are recorded: The test indicates tuberculosis in 96% of those who have it and in 2% of those who do not. Based on these results, what is the probability of a randomly chosen person having tuberculosis, given that the skin test indicates the disease?

Solution: A tree diagram is shown below that represents this situation.

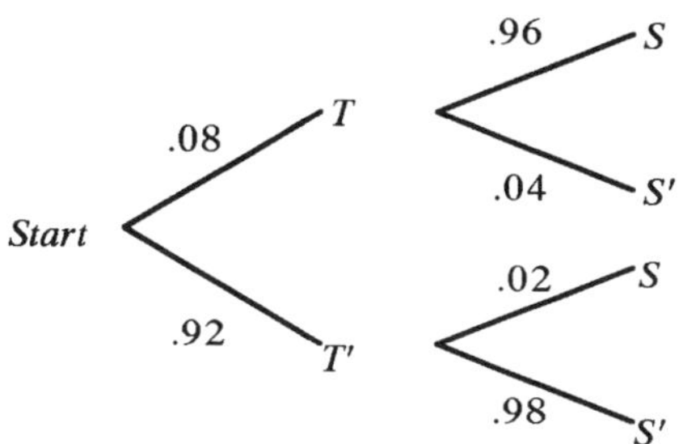

T = Tuberculosis
T' = No tuberculosis
S = Skin test indicates tuberculosis
S' = Skin test indicates no tuberculosis

We are interested in finding P(T|S). This is given by:

$$P(T|S) = \frac{P(T \cap S)}{P(T \cap S) + P(T' \cap S)} = \frac{.08 \times .96}{.08 \times .96 + .92 \times .02} = .81$$

To calculate this enter the following into the calculator:

(.08 × .96) ÷ (.08 × .96 + .92 × .02).

EXAMPLE 5 — *TEXTBOOK SECTION 6-7*

A spinner device is numbered from 1 to 6 and each of the six numbers has a probability of occurring according to the chart below. Any winning bet of $1 will pay $5. A player always bets $1 on the 5. If the spinner lands on the 5 the player wins $4 plus gets the $1 bet back; otherwise, the $1 bet is lost. What is the expected value of the game to this player (long-run average gain or loss per game)? Use two decimal place accuracy. Is this the best number for this player to choose?

Numbers	*1*	*2*	*3*	*4*	*5*	*6*
Probability	$\frac{1}{21}$	$\frac{2}{21}$	$\frac{3}{21}$	$\frac{4}{21}$	$\frac{5}{21}$	$\frac{6}{21}$

Solution: The payoff table for this player is:

x_i	$4	-$1
p_i	$\frac{5}{21}$	$\frac{16}{21}$

To calculate the expected value enter: 4×(5÷21) -1×(16÷21). The result is .19. In the long run this player will win $.19 each game on the average.

Use the replay key to calculate the other possibilities:

Number bet on	*1*	*2*	*3*	*4*	*5*	*6*
Expected value	-.76	-.52	-.29	-.05	.19	.43

The best choice for this player would be to bet on the 6 each game. In the long run this player would then win $.43 each game on the average.

Exercise Set 6

1. Solve Problems 1-12 of Textbook Exercise 6-2.

2. Solve Problems 13-20 of Textbook Exercise 6-2.

3. Consider a group of 18 people.

   (A) In how many ways can 4 be chosen for one committee and 3 be chosen for a second committee? A person can serve on both committees.

   (B) How would the result from Part (A) be affected if there were 5 people chosen for the first committee and 3 for the second committee?

   (C) How would the result from Part (A) be affected if there were only 15 people?

   (D) How would the result from Part (B) be affected if there were only 15 people?

4. From a group of 20 books, in how many ways can 5 be chosen for one bookrack and 10 be chosen for a second bookrack and 3 be chosen for a third bookrack. How would this result be affected if there were 25 books to start with?

5. Twenty-five paintings are available for purchase at an art show. The paintings are numbered from 0 to 24. The artist wishes to choose 5 randomly to sell for charity. What are the painting numbers that he should choose?

6. A group of 30 people are meeting. We want to randomly choose 10 of these people to work on a specific task. If the people are numbered from 1 to 30, who should be chosen?

7. Repeat Example 3 above if a third category, Undecided (U), was added to the vote a person would cast where the Undecided (U) category had 102 Hourly, 87 Salary, and 187 Salary+Bonus votes.

8. Repeat Example 1 if the company did not have a Salary+Bonus category but did have the third category of Undecided (U) from Problem 1.

9. Solve Problems 7-16 of Textbook Exercise 6-6.

10. Solve Problems 17-32 of Textbook Exercise 6-6.

11. A company's brand (X) has 22% of the market. A market research firm finds that if a person currently uses brand X, the probability is .61 that he or she will buy it the next time (represented by X_c). On the other hand, if a person currently does not use X (represented by X_c'), the probability is .53 that he or she will switch to X the next time (represented by X_n).

    (A) Draw a tree diagram representing this situation.

    (B) What is the probability that a person is presently using brand X if he/she purchases brand X on the next purchase?

12. Repeat Problem 3 if the company only had 15% of the market.

13. Solve Problems 1-12 of Textbook Exercise 6-7.

14. Solve Problems 15-20 of Textbook Exercise 6-7.

15. A gaming device is numbered from 1 to 10 and each of the ten numbers has a probability of occurring according to the chart below. If the device lands on the chosen number the player wins $8 and gets the bet back; otherwise, the $2 bet is lost. Suppose a player always bets $2 on the 5.

Numbers	*1*	2	*3*	*4*	*5*	*6*	*7*	*8*	*9*	*10*
Probability	.05	.05	.06	.06	.21	.21	.08	.08	.10	.10

    (A) What is the expected value of the game to this player (long-run average gain or loss per game)? Use two decimal place accuracy.

    (B) Is this the best number for this player to choose?

17. Repeat Problem 15 if the bet is $4 and the player wins $5 and gets the bet back.

18. Do Textbook Chapter 6 Group Activity. Use the calculator where appropriate.

Chapter 7

Data Description and Probability Distributions

This chapter contains examples using the calculator to illustrate:

- Graphing quantitative data
- Calculating measures of central tendency
- Calculating measures of dispersion
- Calculating the probability of x successes in n Bernoulli Trials

The graphing calculator features used are:

- Statistical functions
- Storing an expression
- Evaluating an expression
- Accumulating a sum in a memory location
- Answer key

EXAMPLE 1 **TEXTBOOK SECTION 7-1**

Given the data to the right.

(A) Create a bar graph for the data shown in the table at the right.

(B) Create a broken-line graph for the data.

US. Public Debt	
Year	Billions of Dollars
1950	$ 256.1
1960	284.1
1970	370.1
1980	907.7
1990	3233.3

Solution (A): Assign a number to each year and express the debt as a whole number as shown below:

Year	1950	1960	1970	1980	1990
x	1	2	3	4	5
y	2561	2841	3701	9077	32333

Clear the functions from the function list and any drawings (see Appendix Section A-6, B-6, or C-6 of this manual). Clear the statistical registers in the calculator (see Appendix Section A-17, B-17, or C-17 of this manual).

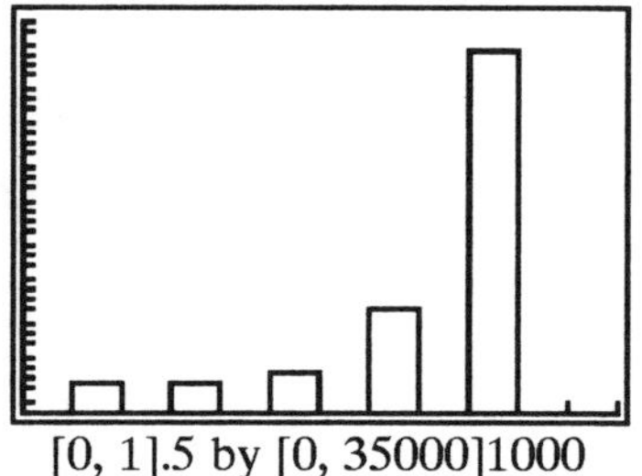

[0, 1].5 by [0, 35000]1000

Store the data using the number as the x variable and the amount as the y variable. Set the viewing window variables at [0, 6].5 by [0, 35000]1000. The bars will be .5 in width and there will be a gap of .5 between bars because an Xscl of .5 will graph data as $0 \le x < .5$, $.5 \le x < 1$, $1 \le x < 1.5$, etc. and there is no data for $.5 \le x < 1$, $1.5 \le x < 2$, etc. The center of the bars is not at the x value, but is at the center of the bar, at $x+.25$

Solution (B): Since the data is already sorted, we do not have to do this step. Clear the drawings (see Appendix Section A-6, B-6 or C-6 of this manual). If the data is not sorted on the x variable, it must be done before the broken-line graph is drawn. Now use the xy-line option on the statistical draw menu to get the graph at the right. See Appendix Section A-16, B-16, or C-16 Example 2(E) of this manual. The line segments are drawn from (1, 2561) to (2, 2841) to (3, 3701) to (4, 9077) to (5, 32333) which are the top left corners of the bars shown in Solution (A) above.

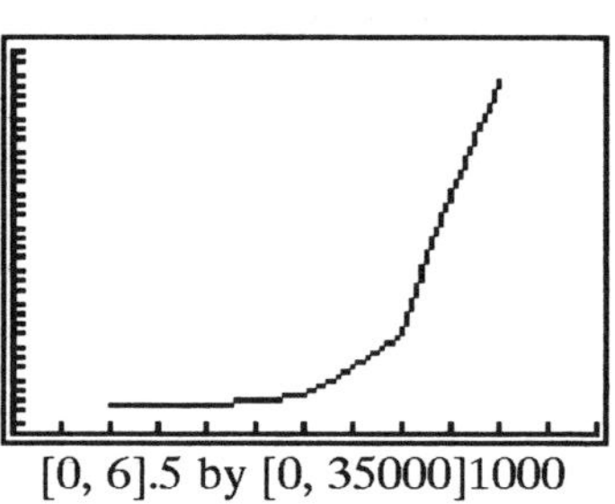

[0, 6].5 by [0, 35000]1000

EXAMPLE 2 ***TEXTBOOK SECTION 7-1***

The following table shows price-earnings ratios of common stocks chosen at random from the New York Stock Exchange:

Price-Earnings (PE) Ratios

7	11	6	6	10	6	31	28	13	19
6	18	9	7	5	5	9	8	10	6
10	3	4	6	7	9	9	19	7	9
17	33	17	12	7	5	7	10	7	9
18	17	4	6	11	13	7	6	10	7

Construct a histogram using 0 as the lower limit, 40 as the upper limit, and class intervals of 5. Use the arrow keys on your calculator to move the cursor and explore the graph. Where do you think the "center" of the data is?

Solution: (See Appendix Section A-16, B-16, or C-16 of this manual on how to enter data and get a histogram for your calculator.) Set the viewing window variables to [0,40]5 by [-5,30]5. The result is shown to the right. Remember that the Xscl is the class width.

[0,40]5 by [-10,30]5

The "center" of the data is usually thought of as being the balance point or the mean. The cursor shown on the graph to the right is probably approximately where you placed your cursor. (Now you see the reason for setting the limits for the *y* axis to start below zero. This gives room on the graph to see the cursor and the coordinates of its location. The coordinates are not shown on the graph above because each calculator model will have different coordinates.)

EXAMPLE 3 — ***TEXTBOOK SECTION 7-2***

Given the data below (see Textbook Section 7-2):

(A) Create a frequency polygon for this data.

(B) Create a relative cumulative frequency ogive for this data.

Class Interval	Midpoint x	Frequency y	Relative Cumulative Frequency
299.5-349.5	324.5	1	.01
349.5-399.5	374.5	2	.03
399.5-449.5	424.5	5	.08
449.5-499.5	474.5	10	.18
499.5-549.5	524.5	21	.39
549.5-599.5	574.5	20	.59
599.5-649.5	624.5	19	.78
649.5-699.5	674.5	11	.89
699.5-749.5	724.5	7	.96
749.5-799.5	774.5	4	1.00

Solution (A): Add a line to the table both at the beginning and at the end.

249.5-299.5	274.5	0	.00
799.5-849.5	824.5	0	1.00

Store the data in the calculator so that x=midpoint and y=frequency.

Sort the data on x. Set the viewing window variables at [250, 850]50 by [0, 25]1. Draw the xy-line. The TI-82 will show the dots if you draw the scatter graph first before the xy-line. (See Appendix Section A-16, B-16, or C-16 of this manual.)

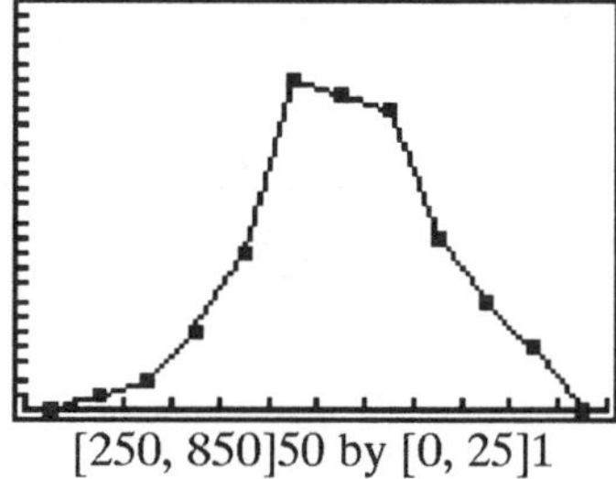

[250, 850]50 by [0, 25]1

Solution (B): Add a line at the beginning of the original table:

249.5-299.5	274.5	0	.00

The calculator needs to have a whole number for the y variable in order to draw the xy-line. Change the relative cumulative frequency to whole numbers by multiplying each by a number, say 100. Store the data in the calculator so that x=upper class boundary and y=whole number representing the relative cumulative frequency. Set the viewing window variables at [250, 850]50 by [0, 150]10. Draw the xy-line. The TI-82 will show the dots if you draw the scatter graph first before the xy-line.

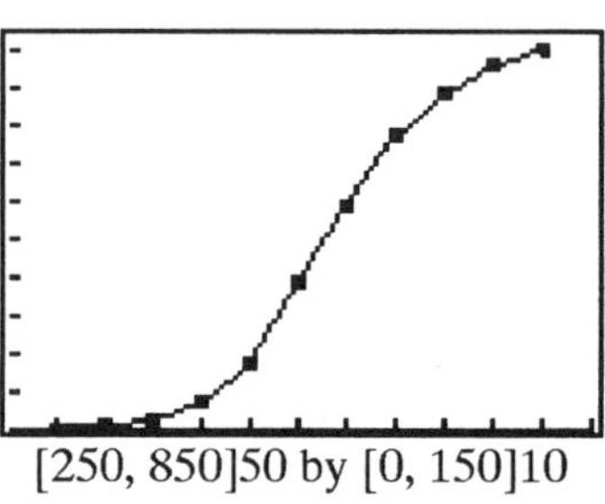

[250, 850]50 by [0, 150]10

EXAMPLE 4 ***TEXTBOOK SECTION 7-3 & 7-4***

Find the mean and standard deviation to two decimal places for the data in Example 2 above.

Solution: Store the data in the calculator with x=the data value and y=the frequency (this is 1 if entering each piece of data separately). (See Appendix Section A-16, B-16, or C-16 of this manual on how to get the one-variable statistics on your calculator.) The results are: $\bar{x}$ = 10.52 and s = 6.63. Since this is a sample of data, we use the sample standard deviation and not population standard deviation.

EXAMPLE 5 ***TEXTBOOK SECTION 7-3 & 7-4***

Find the mean and standard deviation to two decimal places for the data in Example 3 above.

Solution: Store the data using the midpoint of the class as the x variable and the frequency as the y variable as was done in Example 3(A) above. (See Appendix Section A-16, B-16, or C-16 of this manual on how to get the one-variable statistics on your calculator.) The results are: $\bar{x}$ = 579 and s = 94.31. Since this is a sample of data, we use the sample standard deviation and not population standard deviation.

EXAMPLE 6 ***TEXTBOOK SECTION 7-3***

(A) Find the median and mode of the data in Example 2 above.
(B) Find the median and mode of the data in Example 3 above.

Solution (A): Clear the statistical registers in the calculator and enter the data as was done in Example 2 above.

TI-82

The TI-82 calculator has a built-in function that will find the median for you. See Appendix Section B-16 of this manual. The median function is on the 1-Var Stats list. (Use the down arrow to see the last part of the list.)

TI-81 and TI-85

Use the calculator to sort the data on the x variable. (See Appendix Section A-16 or C-16 of this manual.) There are 50 pieces of data so the median value is between the 25th and 26th value. Use the arrow keys while editing the data to find the values in the 25th and 26th position. Take the average of these two values. The median is 9.

The mode is 7.

Solution (B): Clear the statistical registers in the calculator and enter the data as was done in Example 3(A) above.

TI-82

The TI-82 calculator has a built-in function that will find the median for you. See Appendix Section B-16 of this manual. The median function is on the 1-Var Stats list. (Use the down arrow to see the last part of the list.) The median is listed is 574.5. This is the midpoint of the class of the 50.5th value. However, if we want the number that is 11.5/20ths into the class, we must use the method shown for the TI-81 and TI-85 calculators below.

TI-81 and TI-85

Use the calculator to sort the data on the x variable. (See Appendix Section A-16 or C-16 of this manual.) There are 100 pieces of data so the median value is between the 50th and 51st value. Use the arrow keys in the edit mode and count down the frequencies (the y variable) until you find the class having the 50th piece of data. This is the sixth class. Since there are 39 numbers before this class, we need to find the number that is 11 and is 12 into the sixth class and average them. We find the median by taking the fraction of the class width and adding it on to the lower boundary of the class: $\frac{11.5}{\text{number in class}}$ x (class width) + (lower class boundary) = $\frac{11.5}{20}$ x 50 + 549.5 = 578.25.

Hence the median is 578.25.

The modal class has midpoint 524.5. This can be observed in the tabled data or by inspecting the histogram with range [249.5, 824.5]50 by [-5, 25]1.

EXAMPLE 7 ***TEXTBOOK SECTION 7-5***

The probability of recovering after a particular type of operation is .95. Find the probability that at least 5 of the next 50 patients having this operation will not recover.

Solution: At least 5 not recovering is the same as at most 45 will recover. So we want to find the probability of at least five patients not recovering is equivalent to less than 45 patients recovering.

$$\begin{aligned} P(X\le 45) &= P(1)+P(2)+\ldots+P(45) \\ &= 1 - P(X>45) \\ &= 1 - P(X\ge 46) \\ &= 1 - [P(46)+P(47)+P(48)+P(49)+P(50)] \end{aligned}$$

We need to find $P(46)+P(47)+P(48)+P(49)+P(50)$ and subtract the result from 1.

This is a binomial distribution situation with $n = 50$, $p = .95$, and $q = .05$ when a recovery is considered a success. The probability of x people recovering is given by: $P(x) = C_{50,x}\,(.95)^x(.05)^{50-x}$. Store this expression in the calculator as: (50 **nCr** X) (.95 ^ X) .05 ^ (50 - X) where nCr is found on the probability menu (see Appendix Section A-11, B-11, or C-11 of this manual). Store 46 for X and evaluate the expression. Store the result in memory M. Store 47 for X and reevaluate the expression. Add the result to the contents of memory M and store again as memory M. Continue in this manner until the sum of the probabilities is in memory M.

(50 nCr X) (.95 ^ X) .05 ^ (50 - X)
46 → X
Recall the expression and evaluate.
Ans → M — Result is .136

47 → X
Recall the expression and evaluate. — Result is .220
Ans + M → M — Result is .356

48 → X
Recall the expression and evaluate. — Result is .261
Ans + M → M — Result is .617

49 → X
Recall the expression and evaluate. — Result is .202
Ans + M → M — Result is .819

50 → X
Recall the expression and evaluate. — Result is .077
Ans + M → M — Result is .896

Find 1 - M. The result is .104. Thus the probability that at least 5 people out of 50 do not survive the operation is .104 or a little more than 1 chance in 10.

EXAMPLE 8 ***TEXTBOOK SECTION 7-6***

Given the normal probability distribution having mean 15.4 and standard deviation 2.3:

(A) Graph using [6, 24]1 by [0, .3].1.

(B) Find the value of c so that $P(9.2 \le x \le c) = .8534$.

(C) Conjecture, then check by graphing, the location and shape of a normal probability distribution having mean 17 and standard deviation 1.5.

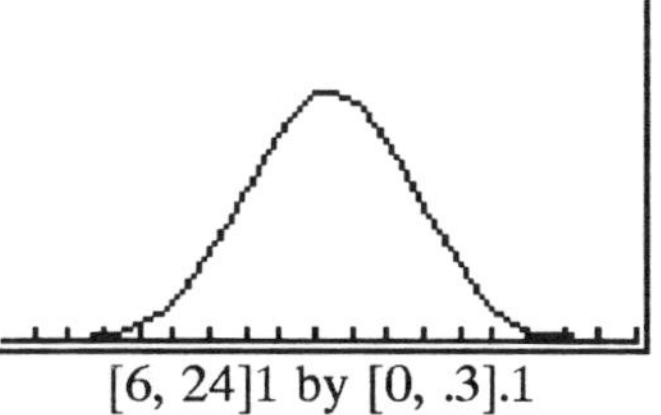
[6, 24]1 by [0, .3].1

Solution (A): The formula is given in the footnote to Textbook Section 7-6. For this situation it is:

$$f(x) = \frac{1}{2.3\sqrt{2\pi}} e^{-\frac{(x - 15.4)^2}{2(2.3^2)}}$$

. Store this in the calculator as:

1 ÷ (2.3 x √ (2 x π)) x [e^] (- ((X - 15.4) ^ 2) ÷ (2 x 2.3 ^ 2)). (Recall that [e^] is found using [2nd] [LN].) Set the viewing window variables to [6, 24]1 by [0, .3].1. Graph.

Solution (B): The z value for x=9.2 is $z = \frac{9.2\text{-}15.4}{2.3}$ = -2.70. From the table we find the area under the curve for z between -2.70 and 0 is .4965. Hence we know that the x value we want is on the right side of the mean which is 15.4. .8534-.4965=.3569 is the area on the right side of x=15.4. This corresponds to a z of .98.

The c value is found by solving $.98 = \frac{c - 15.4}{2.3}$. c = 17.65.

Solution (C): The graph will be shifted to the right 1.6 units and will be taller and skinnier since the standard deviation is smaller.

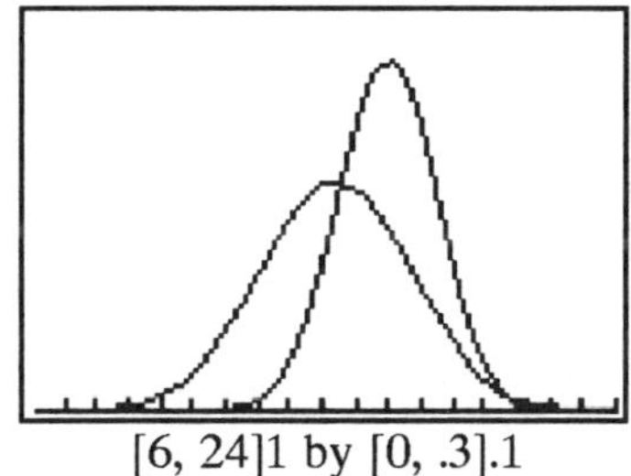
[6, 24]1 by [0, .3].1

Exercise Set 7

1. Solve Problems 1-2 of Textbook Exercise 7-1.

2. Solve Problem 4 of Textbook Exercise 7-1. Use the coding 1=1990 former USSR, 1.3=1991 former USSR, 2=1990 Japan, 2.3=1991 Japan, etc. Set the viewing window variables at [0, 7].3 by [0, 15000]1000.

3. Solve Problem 5 of Textbook Exercise 7-1.

4. Solve Problem 7 of Textbook Exercise 7-1.

5. Solve Problems 1-4 of Textbook Exercise 7-2.

6. Solve Problems 5-7 of Textbook Exercise 7-2.

7. The following table shows scores of students on a 35 question true/false test.

Scores on True/False test

27	31	16	26	10	16	31	28	13	19	16
28	29	17	25	25	29	18	10	26	20	13
24	26	27	19	29	19	17	19	27	33	17
12	27	25	17	10	7	19	18	27	34	26
11	13	17	26	10	27					

Construct a histogram using 0 as the lower limit, 35 as the upper limit, and class intervals of 5.

Use the arrow keys on your calculator to move the cursor and explore the graph. Where do you think the "center" of the data is?

8. Suppose the first column of scores in Problem 7 were all perfect papers with scores of 35. Repeat Problem 1 and discuss how this change of data effects the histogram.

9. Solve Problems 1-12 of Textbook Exercise 7-3.

10. Solve Problems 13-22 of Textbook Exercise 7-3.

11. Find the mean and standard deviation of the sample data in Problem 7.

12. Find the mean and standard deviation of the sample data in Problem 8.

13. Solve Problems 1-4 of Textbook Exercise 7-4.

14. Solve Problems 5-14 of Textbook Exercise 7-4.

15. Solve Problems 1-16 of Textbook Exercise 7-5.

16. Solve Problems 27-34 of Textbook Exercise 7-5.

17. Solve Problems 37-40 of Textbook Exercise 7-5.

13. Repeat Example 7 but find the probability that 3, 4 or 5 will not recover.

14. Repeat Example 7 but change the probability of recovering to .97.

15. Solve Problems 17-24 of Textbook Section 7-6.

16. Solve Problems 35-42 of Textbook Section 7-6.

17. Solve Problems 43-46 of Textbook Section 7-6.

18. Do Textbook Chapter 7 Group Activity. Use the calculator where appropriate.

Chapter 8

Markov Chains

This chapter contains examples using the calculator to illustrate:

- Finding next state and stationary state matrices in Markov chains
- Finding powers of the transition matrix
- Determining if a transition matrix is regular
- Finding a limiting matrix of a transition matrix

The graphing calculator features used are:

- Matrices
    - Storing
    - Multiplying by a constant
    - Recalling a single element
    - Finding the matrix product of two matrices
    - Finding the inverse
    - Using the inverse matrix to solve a system of equations
- Ans key

EXAMPLE 1 ***TEXTBOOK SECTION 8-1, 8-2, and 8-3***

A given plant species has red, pink, or white flowers according to the genotypes *RR*, *RW*, and *WW*, respectively. If each of these genotypes is crossed with a pink-flowering plant (genotype *RW*), then the transition matrix *P* is:

		Next generation		
		Red	Pink	White
$P =$ **This generation**	Red	.50	.50	.00
	Pink	.25	.50	.25
	White	.00	.50	.50

Assuming the plants of each generation are crossed only with pink plants to produce the next generation:

(A) Find, to two decimal places, the percentage of each color plant after four generations and after 10 generations if the initial generation is all red plants.

(B) Show that, regardless of the makeup of the first generation, the genotype composition will eventually stabilize at 25% red, 50% pink, and 25% white.

(C) Repeat Parts (A) and (B) if the initial generation is all white plants.

(D) Conjecture, then check, what the fourth generation, tenth generation, and stationary matrix will be if the initial matrix is $S_0 = [0 \quad 1 \quad 0]$.

(E) Is P a regular matrix?

(F) Find the limiting matrix $\bar{P}$.

Solution (A): We need to find the fifth state matrix with initial matrix of $S_0 =$ $[1 \quad 0 \quad 0]$. Store the transition matrix P in the calculator. The fourth state matrix is found by calculating $S_0P^4 = [1 \quad 0 \quad 0]\begin{bmatrix} .50 & .50 & .00 \\ .25 & .50 & .25 \\ .00 & .50 & .50 \end{bmatrix}^4$. To do this, store the 3×3 matrix $P = \begin{bmatrix} .50 & .50 & .00 \\ .25 & .50 & .25 \\ .00 & .50 & .50 \end{bmatrix}$ as matrix A in the calculator. Also, store the 1×3 matrix $S_0 = [1 \quad 0 \quad 0]$ as matrix B in the calculator. (See Appendix Section A-12, B-12, or C-12 of this manual.) Find the fourth power of A and store it as matrix C. This result is $\begin{bmatrix} .28125 & .50000 & .21875 \\ .25000 & .50000 & .25000 \\ .21875 & .50000 & .28125 \end{bmatrix}$. Now multiply the initial state matrix by matrix C. The result is $[.28125 \quad .50000 \quad .21875]$. An alternative method would be to find BA^4 directly using matrix multiplication without finding A^4 first.

We also need to find $S_0P^{10} = BA^{10}$. The result is $[.25049 \quad .50000 \quad .24951]$. We will see that this is very close to the stationary matrix.

Solution (B): We need to find the stationary matrix. This matrix is found by solving the matrix equation:

$$[m \quad n \quad p]\begin{bmatrix} .50 & .50 & .00 \\ .25 & .50 & .25 \\ .00 & .50 & .50 \end{bmatrix} = [m \quad n \quad p].$$ This is the same as

$$[m \quad n \quad p]\begin{bmatrix} .50 & .50 & .00 \\ .25 & .50 & .25 \\ .00 & .50 & .50 \end{bmatrix} = [m \quad n \quad p]\begin{bmatrix} 1 & 0 & 0 \\ 0 & 1 & 0 \\ 0 & 0 & 1 \end{bmatrix}$$

$$[m \;\; n \;\; p]\left\{\begin{bmatrix} .50 & .50 & .00 \\ .25 & .50 & .25 \\ .00 & .50 & .50 \end{bmatrix} - \begin{bmatrix} 1 & 0 & 0 \\ 0 & 1 & 0 \\ 0 & 0 & 1 \end{bmatrix}\right\} = [\,0 \;\; 0 \;\; 0\,]$$

$$[m \;\; n \;\; p]\begin{bmatrix} -.50 & .50 & .00 \\ .25 & -.50 & .25 \\ .00 & .50 & -.50 \end{bmatrix} = [\,0 \;\; 0 \;\; 0\,].$$ This result is equivalent to

$$\begin{aligned} -.50m + .25n \phantom{{}+.50p} &= 0 \\ .50m - .50n + .50p &= 0. \\ .25n - .50p &= 0 \end{aligned}$$ As is shown in the text, this is a dependent system.

Only

two of these equations can be used. The third equation to use in finding the stationary matrix is $m+n+p=1$. Hence we need

to solve the matrix equation $\begin{bmatrix} -.50 & .25 & .00 \\ .50 & -.50 & .50 \\ 1.00 & 1.00 & 1.00 \end{bmatrix}\begin{bmatrix} m \\ n \\ p \end{bmatrix} = \begin{bmatrix} 0 \\ 0 \\ 1 \end{bmatrix}$. Enter the 3x3 coefficient matrix into the calculator as matrix A and find its inverse. This result is

$\begin{bmatrix} -2.00 & -.50 & .25 \\ .00 & -1.00 & .50 \\ 2.00 & 1.50 & .25 \end{bmatrix}$. Store this as matrix C. Store the constant 3×1 matrix as matrix B in the calculator. Multiply the inverse matrix by the constant matrix, CB. The result is $\begin{bmatrix} .25 \\ .50 \\ .25 \end{bmatrix}$. The stationary matrix is then [.25 .50 .25] to two decimal place accuracy. The expected genotype composition will eventually stabilize at 25% red, 50% pink, and 25% white.

Solution (C): We need to find the product

$S_0P^4 = BA^4 = [\,0 \;\; 0 \;\; 1\,]\begin{bmatrix} .50 & .50 & .00 \\ .25 & .50 & .25 \\ .00 & .50 & .50 \end{bmatrix}^4$. Store the 1×3 matrix S_0 as matrix B in the calculator. Store the 3×3 matrix P as matrix A in the calculator. Find the matrix product BA^4. The result is [.21875 .50000 .28125].

The tenth state matrix is [.24951 .50000 .25049].

The steady state matrix is the same as found in Part (B) above.

Solution (D): A good conjecture would be that the fourth state matrix would be close to [.24 .52 .24], that the tenth state matrix would be close to the stationary matrix, and that the stationary matrix is the same as in Parts (B) and (C).

As it turns out, $S_0P^4 = BA^4 = [0 \quad 1 \quad 0]\begin{bmatrix} .50 & .50 & .00 \\ .25 & .50 & .25 \\ .00 & .50 & .50 \end{bmatrix}^4 = [\,.25 \ .50 \ .25\,]$ = stationary matrix.

Solution (E): P is a regular matrix since A^2 has all positive elements.

Solution (F): Find successive powers of P to find $\bar{P}$. Do this on the calculator using matrix A since this is where we stored the elements of matrix P.

$\bar{P} = \begin{bmatrix} .25 & .50 & .25 \\ .25 & .50 & .25 \\ .25 & .50 & .25 \end{bmatrix}$. We need to find A^{34} (A^{40} on the TI-85) to see the limiting matrix. An easy way to do this is to recall matrix A. Then enter [Ans] × [A]. Pressing the [ENTER] key repeatedly will continue to perform this multiplication. Count the number of times you push the [ENTER] key.

Exercise Set 8

1. Solve Problems 1-8 of Textbook Exercise 8-1.

2. Solve Problems 29-32 of Textbook Exercise 8-1.

3. Solve Problems 35-36 of Textbook Exercise 8-1.

4. Solve Problems 39-42 of Textbook Exercise 8-1.

5. A company's brand X has 22% of the market. A market research firm finds that if a person uses brand X now, represented by X_0, the probability is .61 that he or she will buy it the next time, represented by X_1. On the other hand, if a person does not use brand X now (represented by X_0'), the probability is .53 that he or she will switch to brand X the next time, represented by X_1'.

   (A) Draw a probability tree diagram representing this situation.

   (B) What is the probability that a person is presently using brand X, represented by X_0, if he/she purchases brand X on the next purchase, represented by X_1?

6. Repeat Problem 5 if the company only had 15% of the market.

7. Refer to Problem 5 above. (A) What share of the market will the company have after 5 purchases? (B) What share of the market will the company have in the long-run?

8. Repeat Problem 5 if in Problem 5 the probability is .75 that a person presently using brand X, represented by X_0, will purchase it the next time.

9. Repeat Problem 5 if the company introduced a new product Y so that the initial matrix is [.35 .21 .44] and the transition matrix is

$$\begin{array}{r} \\ \\ X_0 \\ Y_0 \\ \textit{neither } X_0 \textit{ nor } Y_0 \end{array} \begin{array}{ccc} X_1 & Y_1 & \textit{neither} \\ & & X_1 \textit{ nor } Y_1 \\ \left[\begin{array}{c} .58 \\ .35 \\ .13 \end{array}\right. & \begin{array}{c} .29 \\ .45 \\ .17 \end{array} & \left.\begin{array}{c} .13 \\ .20 \\ .70 \end{array}\right] \end{array}.$$

10. Repeat Problem 9 where $P(X_1|X_0) = .63$, $P(Y_1|X_0) = .21$, $P(\textit{neither}\ |X_0) = .16$ with the remaining probabilities the same. In other words, the first row of the matrix in Problem 9 is .63, .21, .16.

11. Solve Problems 1-12 of Textbook Exercise 8-2.

12. Solve Problems 13-20 of Textbook Exercise 8-2.

13. Solve Problems 21-24 of Textbook Exercise 8-2.

14. Solve Problems 41-42 of Textbook Exercise 8-2.

15. Solve Problems 17-22 of Textbook Exercise 8-3.

16. Solve Problems 23-28 of Textbook Exercise 8-3.

17. Solve Problems 29-32 of Textbook Exercise 8-3.

18. Do Textbook Chapter 8 Group Activity. Use the calculator where appropriate.

NOTES

Chapter 9

The Derivative

This chapter contains examples using the calculator to illustrate:

- Evaluating functions
- Finding limits using tables
- Finding limits from graphs
- Finding average rate of change
- Finding the instantaneous rate of change
- Finding the derivative using the difference quotient
- Finding the numeric derivative
- Finding an equation of the tangent line

The features of a graphing calculator that are used are:

- Calculation
- Storing a function
- Evaluating a function
- Graphing and setting viewing window variables
- Tracing and zooming

EXAMPLE 1 ***TEXTBOOK SECTION 9-1, 9-3, 9-7***

The revenue (in dollars) from the sale of x plastic planters is given by

$$R(x) = 20x - 0.02x^2 \qquad 0 \le x \le 1000$$

(A) Graph the function.

(B) What is the change in revenue if production is changed from 615 planters to 900 planters? Draw this on the graph.

(C) What is the average change in revenue for this change in production?

(D) Find the instantaneous rate of change of revenue when 615 planters are produced.

(E) Find the slope of the secant line joining $(615, R(615))$ and $(900, R(900))$.

(F) Find an equation of the tangent line to the graph of $R(x)$ at $x = 615$. Graph $R(x)$ and the tangent line on the same set of coordinate axes.

(G) Find the marginal revenue at $x = 615$. Draw the secant line joining $(615, R(615))$ and $(616, R(616))$ using viewing window variables [575, 650]0 by [4500, 4900]0. Draw the tangent line at $x = 615$ on the same set of coordinate axes and compare to the secant line.

Solution (A): Store the function $R(x)$ as Y1 in the calculator as 20 X - .02 X ^ 2. Enter the viewing rectangle variables as [0, 1000]100 by [0, 5500]500. Graph.

Solution (B): Evaluate the function at x=615. The point on the graph has coordinates (615, 4735.5). Evaluate the function at x=900. The point on the graph has coordinates (900, 1800). Draw the line segment between these two points (see Appendix Section A-3, B-3, or C-3 of this manual).

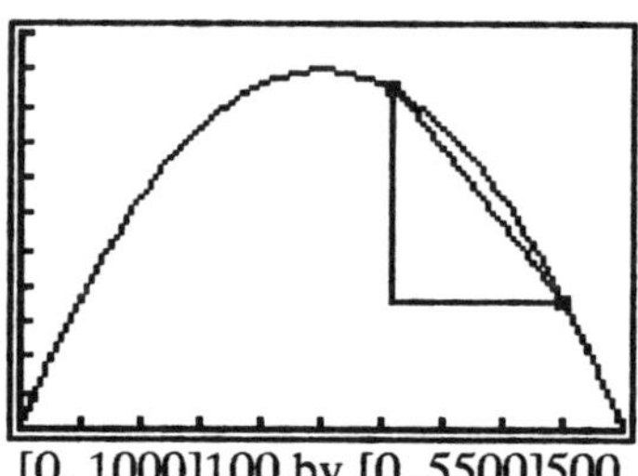
[0, 1000]100 by [0, 5500]500

Now draw the line segments between (615, 4735.5) and (615, 1800) and between (615, 1800) and (900, 1800).

The change in revenue is $R(900) - R(615) = 1800 - 4735.5 = -2935.5$. The negative sign means that the revenue is decreasing as the production increases.

Solution (C): The average change in revenue is found by evaluating $\frac{R(900)-R(615)}{900-615} = \frac{1800 - 4735.5}{900 - 615} = -10.3$. This means that the average change in revenue is \$10.30 per planter when production is increased from 615 to 900 planters.

Solution (D): The instantaneous rate of change is found by evaluating

$$\lim_{h\to 0} \frac{f(615+h) - f(615)}{h} = \lim_{h\to 0} \frac{(20(615+h)-.02(615+h)^2)-(20\text{x}615-.02\text{x}615^2)}{h}$$

$= \lim_{h\to 0} -4.6-.02h = -4.6$. This means that the revenue is \$4735.50 and is decreasing at a rate of \$4.60 per year when the production was 615 planters.

This can also be seen from a table shown below. To calculate the difference quotient, store it as Y1 as ((20 (615 + H) - .02 (615 + H) ^ 2) - (20 x 615 - .02 x 615 ^ 2)) ÷ H . Store a value for H and evaluate Y1. Repeat for all values of H except 0.

h	-0.1	-0.01	-0.001	→	0	←	0.001	0.01	0.1
$\frac{f(615+h) - f(615)}{h}$	-4.598	-4.5998	-4.59998	→	-4.6	←	-4.60002	-4.6002	-4.602

Solution (E): Finding the slope of the secant line is the same as finding the average change in revenue. This was found in Part (C) above. The slope is -10.3.

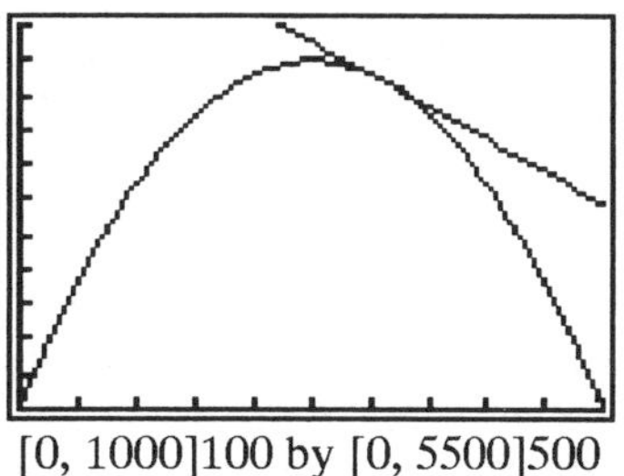
[0, 1000]100 by [0, 5500]500

Solution (F): The slope of the tangent line is the same as the instantaneous rate of change. This was found in Part (D) above. The slope is -4.6. The point on the graph is (615, 4735.5). Using the point-slope form of a line, we have (y - 4735.5) = -4.6(x - 615). Change this to y = 4735.5 - 4.6(x-615) and graph using viewing window variables of [0, 1000]100 by [0, 5500]500.

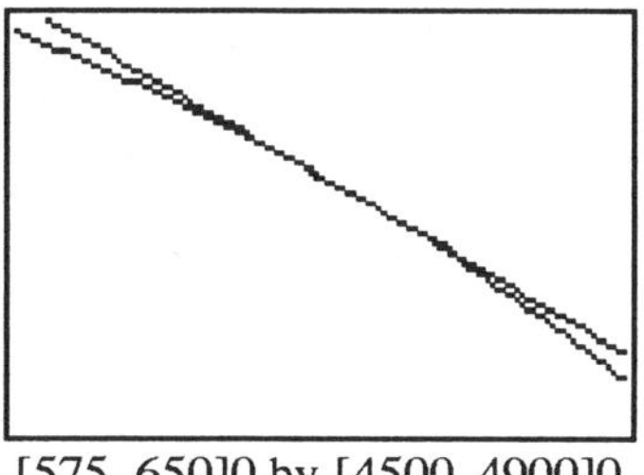
[575, 650]0 by [4500, 4900]0

Solution (G): The marginal revenue is the same as the instantaneous rate of change. This was found in Part (D) and Part (F) above. Enter the function and the tangent line equation in the calculator if they are not already stored. Set the viewing window variables to [600, 620]0 by [4500, 4800]0. Evaluate $R(616)$ to find the coordinates of the second point to use in drawing the secant line. The coordinates are (616, 4730.88). Draw the line segment on the graph. You will not see it drawn because the line segment is so close to the graph itself. (See Appendix Section A-3, B-3, or C-3.) If you change the viewing window variables to [614, 617]0 by [4730, 4736]0 and graph you will see that the tangent line graphs almost entirely on the revenue function. This shows how good an approximation the tangent line is to the revenue function for small changes in x.

EXAMPLE 2 ***TEXTBOOK SECTION 9-1***

A small steel ball dropped from a tower will fall a distance of y feet in x seconds, as given approximately by the formula (from physics) $y = f(x) = 16x^2$. Investigate the average velocity of the ball from x=2 to x=2+h seconds as $h \to 0$. Use four decimal places.

Solution: The average velocity is given by $\dfrac{f(x+h) - f(x)}{h} = \dfrac{16(x+h)^2 - 16x^2}{h}$.

Store this in the calculator as (16 (X + H) ^ 2 - 16 X ^ 2) ÷ H . Store 2 as X and .01 as H. Evaluate the expression. The result is 64.1600. Store .005 as H and evaluate again. Continue storing values for H and evaluating the expression to complete the chart below.

h	0.01	0.005	0.001	0.0001	$\to 0$
$\dfrac{16(x+h)^2 - 16x^2}{h}$	64.1600	64.0800	64.0160	64.0016	$\to 64$

Repeat using negative values of h = -0.01, -0.005, -0.001, and -0.0001. The limit is the same. Hence, the instantaneous velocity of the ball is 64 ft/sec at x=2 seconds.

EXAMPLE 3 ***TEXTBOOK SECTION 9-2***

Given the functions $f(x) = \frac{x^2-4}{x-2}$. Examine the graph of $f(x)$ for values of x near 2 to determine $\lim_{x \to 2} f(x)$.

Solution: This function was graphed in Chapter 2 Example 1 of this manual. Graph using [-4.7, 4.8]1 by [-12.8, 12.4]1 on the TI-81, [-4.7, 4.7]1 by [-12.4, 12.4]1 on the TI-82, or [-6.3, 6.3]1 by [-12.4, 12.4]1 on the TI-85. The graph is shown to the right. Note that there is a hole in the graph at x=2.

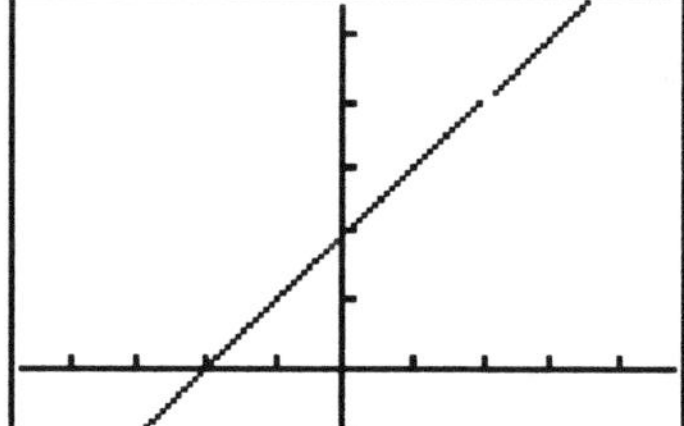

When you use the trace feature of the calculator you see that the y values are close to 4 when x is close to 2. Notice, however, that there is no coordinate for y at the bottom of the screen when the x coordinate is 2. This indicates that the function is not defined for values of 2. $\lim_{x \to 2} f(x) = 4$ but $f(2)$ is not defined.

EXAMPLE 4 ***TEXTBOOK SECTION 9-2***

Given the function $f(x) = \frac{|x-2|}{x-2}$. Investigate the behavior of x near 2 to determine $\lim_{x \to 2} f(x)$.

Solution (A): This function was examined in Chapter 2 Example 4 of this manual. The table and graph are repeated below.

x	1.900	1.990	1.999	$\rightarrow 2 \leftarrow$	2.001	2.010	2.100
$f(x)$	-1	-1	-1 $\rightarrow$?	?$\leftarrow$	1	1	1

We see the that the function values are -1 for values of x less than 2. The function values are 1 for values of x greater than 2.

Since the limit from the left and the limit from the right of 2 are not equal, $f(x)$ does not have a limit at x = 2.

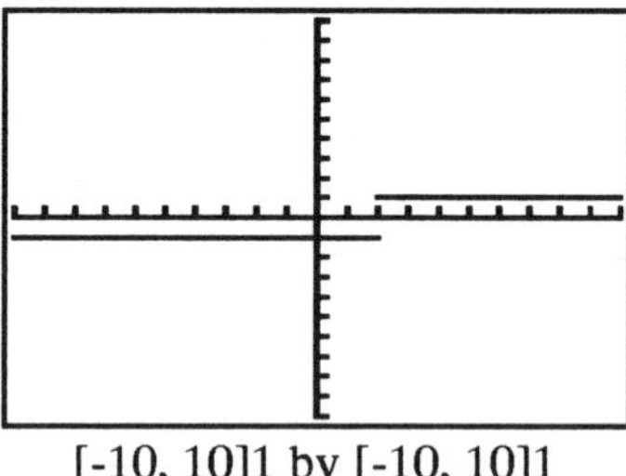

[-10, 10]1 by [-10, 10]1

EXAMPLE 5 ***TEXTBOOK SECTION 9-2***

Investigate the behavior of $f(x) = \dfrac{3x^2 + 5x - 8}{x^2 - 1}$ when x is near to (A) 1 (B) -1 .
Does the limit exist at these values of x ?

Solution: This function was examined in Chapter 2 Example 3 of this manual. The tables and graph are repeated below.

x	.900	.990	.999	→ 1 ←	1.001	1.010	1.100
$f(x)$	5.632	5.513	5.501→ ?		?←5.499	5.488	5.381

x	-1.100	-1.010	-1.001	→ -1 ←	-.999	-.990	-.900
$f(x)$	-47	-497	-4997→ ?		?←5003	503	53

We found that $\lim_{x \to 1} f(x) = 5.5$ and $\lim_{x \to -1} f(x)$ does not exist.

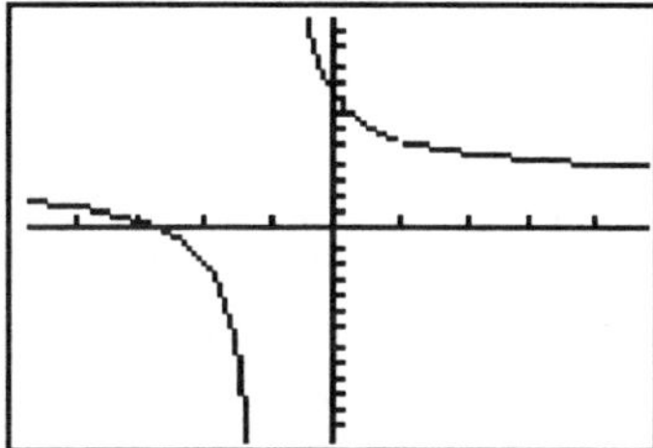

EXAMPLE 6 ***TEXTBOOK SECTION 9-3***

Given $f(x) = \begin{cases} x^2 & \text{if } x \le 3 \\ -2x + 5 & \text{if } x > 3 \end{cases}$. (A) Graph. (B) Determine where f is nondifferentiable.

Solution (A): There is a trick to graphing piecewise-defined functions on the calculator. We can graph the two pieces as separate functions but each piece should be divided by an inequality expression that will have value 1 when the inequality is true but 0 when the inequality is not true. Hence we divide the first piece x^2 by $(x \le 3)$ and the second piece $-2x + 5$ by $(x>3)$. (See Appendix Section A-6, B-6, or C-6 of this manual.)

Store X ^ 2 ÷ (X ≤ 3) for the first function and store (- 2 X + 5) ÷ (X > 3) as the second function. Graph using [-10, 10]1 by [-10, 10]1.

Solution (B): There is a break in the graph at $x = 3$. Hence the function is nondifferentiable at this point.

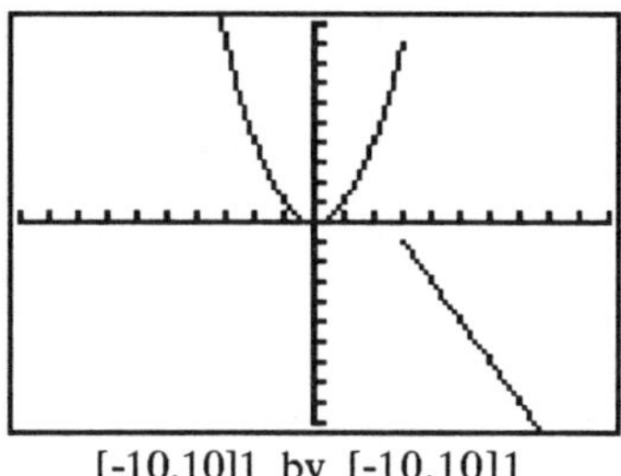
[-10,10]1 by [-10,10]1

EXAMPLE 7 ***TEXTBOOK SECTION 9-3***

Find $f'(x)$ for $f(x) = \dfrac{3x^2 + 5x - 8}{x^2 - 1}$ at $x=1$.

Solution: The calculator will find the numeric derivative of a function. (See Appendix Section A-15, B-15, or C-15 of this manual.) Enter the function in the function list as Y1 as (3 X ^ 2 + 5 X - 8) ÷ (X ^ 2 - 1). Follow the procedure for your particular calculator to find the numeric derivative at $x = 1$. The derivative is -1.25.

EXAMPLE 8 ***TEXTBOOK SECTION 9-2***

Show that $f(x) = \sqrt[3]{(x-2)^2}$ is not differentiable at $x=2$ using a table. Sketch the graph on paper and draw two tangent lines on the graph for $x = 2.2$ and $x = 3$.

Solution: The derivative of this function is $\lim_{h \to 0} \dfrac{\sqrt[3]{((x+h)-2)^2} - \sqrt[3]{(x-2)^2}}{h}$. Store the expression in the calculator as (((X + H - 2) ^ 2) ^ (1 ÷ 3) - ((X - 2) ^ 2) ^ (1 ÷ 3)) ÷ H . Store 2 as X. Evaluate the expression at H = .01, .005, .001, .0001, and .00001 to see that the values of the derivative continue to increase without bound (see table below). Since the limit does not exist the derivative does not exist and the function is not differentiable at $x=2$.

h	.01	.005	.001	.0001	.00001	→ 0
difference quotient	4.64	5.85	10	21.54	46.42	→ does not exist

The graph of the function with the two tangent lines at $x = 2.2$ and $x = 3$ is shown below. We see that the tangent lines get more and more vertical as x gets closer and closer to 2 but a little greater than 2. In other words, $2+h \to 2$ as $h \to 0$.

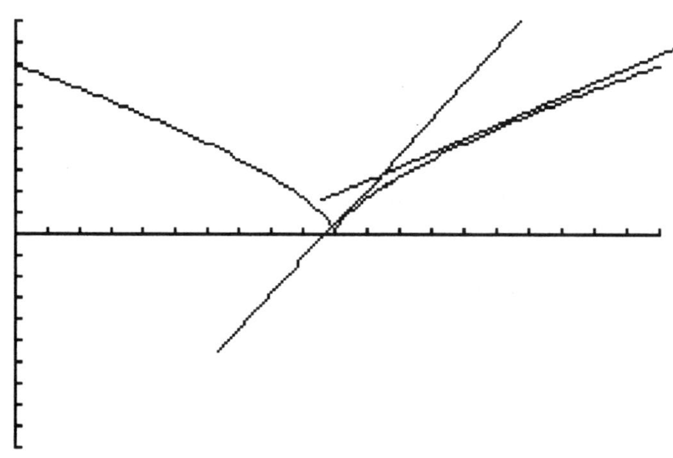

EXAMPLE 9 ***TEXTBOOK SECTION 9-7***

Refer to Example 1 of this chapter. Find the marginal revenue at the production rates of 300, 500, and 800 planters. Interpret each case.

Solution: Enter the function in the calculator as 20 X - .02 X ^ 2. Evaluate the numeric derivative at 300, 500, and 800 planters. The results are: 8, 0, and -12.

This means that the approximate increase revenue from producing 301 items is 8 more than when producing 300 items. There is approximately no change in revenue if the production rate is increased from 500 to 501 planters. The revenue decreases approximately $12 if the production rate increases 1 planter when currently producing 800 planters.

EXAMPLE 10 ***TEXTBOOK SECTION 9-7***

Refer to Example 1. The revenue (in dollars) from the sale of x plastic planters is given by $R(x) = 20x - 0.02x^2$ for $0 \leq x \leq 1000$. Suppose the cost function for this situation is given by $C(x) = 80\, e^{.005x}$ for $0 \leq x \leq 1000$.

(A) Graph both functions on the same set of rectangular coordinates.

(B) Find, to two decimal places, the break-even points. Interpret.

Solution (A): Set the viewing window variables at [0, 1000]100 by [0, 5000]500. Enter the function $R(x)$ as Y1 as 20 X - .02 X ^ 2. Enter the function $C(x)$ as Y2 as 80 [e^x] (.005 X) where e^x is a built-in function found above the [LN] key. Graph. See the top graph to the right.

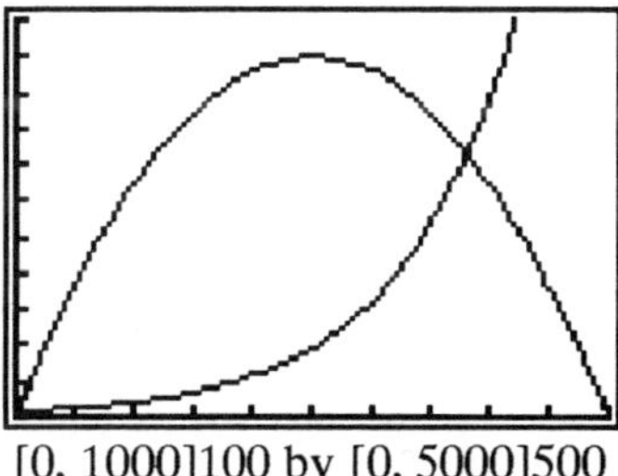

[0, 1000]100 by [0, 5000]500

Solution (B): We want to find where $R(x) = C(x)$ or $20x - 0.02x^2 = 80\, e^{.005x}$. Use trace and zoom, or other calculator methods, to find the coordinates of these points. (See Appendix Section A-7 & A-9, B-7 & B-9, or C-7 & C-9 of this manual on how to zoom and solve equations.) The coordinates of the intersection points of these two graphs are (4.10, 81.66) and (762.53, 3621.58).

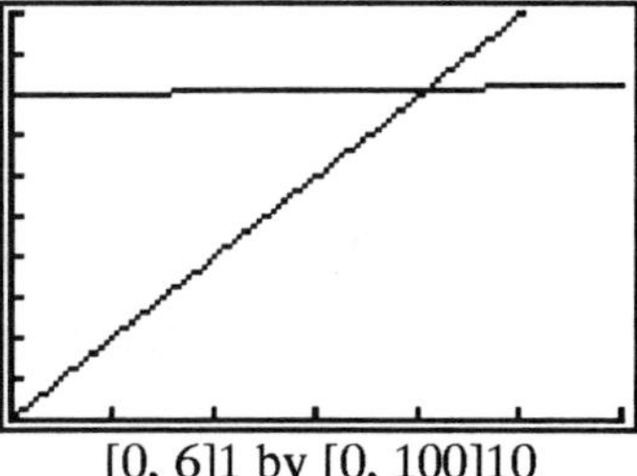

[0, 6]1 by [0, 100]10

The first break-even point shows that the profit will be negative when 4 planters are produced and positive when 5 planters are produced since the break-even point occurs when x=4.10. The revenue and cost are about $80. See the bottom graph to the right.

The second break-even point shows that the profit will be negative when 762 planters are produced and positive when 763 planters are produced. The revenue and cost are about \$3621. See the graph to the right.

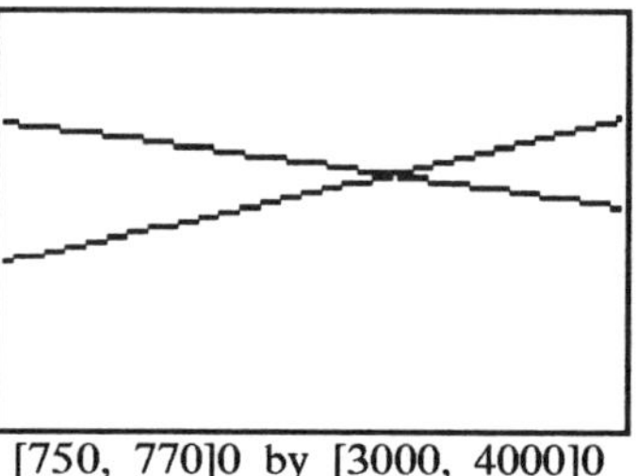

[750, 770]0 by [3000, 4000]0

Exercise Set 9

1. Solve Problems 1-14 of Textbook Exercise 9-1.

2. Solve Problems 15-20 of Textbook Exercise 9-1.

3. Solve Problems 25-52 of Textbook Exercise 9-2 by graphing the function and exploring the graph for x values close to the value where the denominator equals zero. See Examples 3, 4, and 5 above.

4. Solve Problems 53-62 of Textbook Exercise 9-2 by storing the difference quotient in the function list of the calculator and evaluating for h = -0.1, -0.01, -0.001, 0.001, 0.01, and 0.1.

5. Solve Problems 63-66 of Textbook Exercise 9-2 by finding an equation for the tangent line, entering the function and the tangent line expression into the function list of the calculator, setting the viewing window variables, and graphing.

6. Solve Problems 71-72 of Textbook Exercise 9-2.

7. Investigate the behavior of $f(x) = \dfrac{2x^2 - 7x + 5}{x^2 - 2x + 1}$ when x is near 1. Does the limit exist at this value of x ?

8. Investigate the behavior of $f(x) = \dfrac{2x^2 - 3x - 5}{x^2 - 4x + 3}$ when x is near (A) 1 (B) 3. (C) Does the limit exist at these values of x ?

9. Investigate the behavior of $f(x) = \dfrac{x-2}{|x-2|}$ for x near 2. Does the limit exist at $x = 2$? (Compare to Example 4 above. Is this the same function?)

10. Investigate the behavior of $f(x) = \dfrac{|x+3|}{x+3}$ for x near -3. Does the limit exist at $x = -3$?

11. Solve Problems 15-18 of Textbook Exercise 9-3 by first graphing the function. Graph the function and the tangent line on the same set of coordinate axes on the calculator.

12. Solve Problems 19-24 of Textbook Exercise 9-3 by storing the difference quotient in the function list and evaluating. (See Example 2 above.)

13. Solve Problems 33-36 of Textbook Exercise 9-3.

14. Solve Problems 41-44 of Textbook Exercise 9-3.

15. Solve Problems 49-50 of Textbook Exercise 9-3 by graphing the function and investigating using trace and zoom.

16. Solve Problems 55(B) & (C) and 56(B) & (C) of Textbook Example 9-3 by storing the expression in the function list of the calculator and making a table.

17. Graph f and locate all points of discontinuity.

$$f(x) = \begin{cases} 4.2x + 2 & \text{if } x < -.5 \\ x^2 + 3 & \text{if } x = -.5 \\ x^3 - 2 & \text{if } x > -.5 \end{cases}$$

18. Graph f and locate all points of discontinuity.

$$f(x) = \begin{cases} x - 4 & \text{if } x < 3 \\ x^2 - 3 & \text{if } x \geq 3 \end{cases}$$

19. Use graphing techniques to find $\lim_{x\to\infty} \dfrac{4x^3 - 2x + 3}{7x^2 + 3}$.

20. Use graphing techniques to find $\lim_{x\to\infty} \dfrac{7x^2 + x - 5}{5x^4 - 2}$.

21. Show that $f(x) = \sqrt[3]{x - 3} + 1$ is not differentiable at $x = 3$. Sketch the graph on paper and draw a tangent line on the graph for $x = 2.8$.

22. Show that $f(x) = \sqrt[3]{(x + 3)^2}$ is not differentiable at $x = -3$ using a table. Sketch the graph on paper and draw two tangent lines on the graph for $x = -2.8$ and $x = -2$.

23. Solve Problems 13-18 of Textbook Exercise 9-7.

24. Solve Problems 19-20 of Textbook Exercise 9-7.

25. Do Textbook Chapter 9 Group Activity. Use the calculator where appropriate.

NOTES

Chapter 10

Graphing and Optimization

This chapter contains examples using the calculator to illustrate:

- Evaluating algebraic expressions
- Finding limits using tables
- Finding limits from graphs
- Investigating the continuity of a function at a value of x
- Finding the instantaneous velocity
- Finding the derivative using the difference quotient

The features of a graphing calculator that are used are:

- Calculation
- Storing an algebraic expression
- Evaluating an algebraic expression
- Graphing and graph screen limits
- Tracing and zooming

EXAMPLE 1 ***TEXTBOOK SECTION 10-1***

Refer to Chapter 2 Example 2 and Chapter 9 Example 5 of this manual where we investigated the behavior of $f(x) = \dfrac{3x^2 + 5x - 8}{x^2 - 1}$ when x is near to $x=1$, and $x=-1$.

(A) Is the function continuous at these values of x?

(B) Does the function have an indeterminate form at any values of x?

(C) Graph using [-4.7, 4.8]1 by [-12.8, 12.4]1 on the TI-81, [-4.7, 4.7]1 by [-12.4, 12.4]1 on the TI-82, or [-6.3, 6.3]1 by [-12.4, 12.4]1 on the TI-85.

(D) Write the right-hand and left-hand limits at $x=-1$ using limit notation.

(E) Write the equation of the vertical asymptote.

(F) Solve $f(x)<0$.

Solution (A): We found in Chapter 9 Example 5 of this manual that $\lim_{x \to 1} f(x) = 5.5$ but that f(1) is not defined. Hence, the function is not continuous at x=1.

We also found in Example 5 of Chapter 9 of this manual that $\lim_{x \to -1}$ f(x) does not exist. Hence the function is not continuous at $x = -1$.

Solution (B): Yes, when $x = -1$.

Solution (C): We use these values for the viewing window variables because of the number of pixels on the graph screens. By using these settings, the coordinates for x and y at the bottom of the viewing window are in steps of .1 when you use the arrow keys to find screen locations (not trace) of the calculator. This is because the screen on the TI-81 is 96 pixels across by 64 pixels down; on the TI-82 it is 95 pixels across by 63 pixels down; and on the TI-85 the screen is 127 pixels across by 63 pixels down.

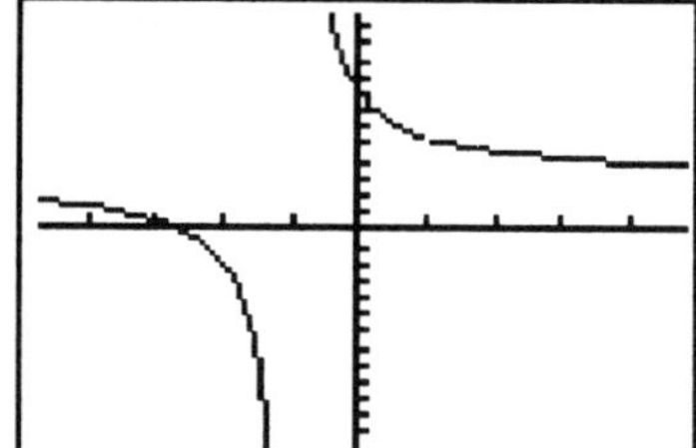

Solution (D): $\lim_{x \to -1^-} f(x) = -\infty$ and $\lim_{x \to -1^+} f(x) = \infty$

Solution (E): The vertical asymptote has equation $x = -1$.

Solution (F): We want to find the values of x where $f(x) < 0$. This occurs when the graph of $f(x)$ is below the x axis. So we first want to find the x intercept. From the graph in Part (C) above, we see that the x intercept is about -2.5. Using trace and zoom or other calculator methods, we find that the x intercept is approximately -2.67, to two decimal places. The exact value is $-\frac{8}{3}$. This can be found by setting $f(x)$=0 and solving for x. Hence the solution to this inequality is $-\frac{8}{3} \le x < -1$.

This problem could also have been solved using a sign chart by first factoring the numerator and denominator of the rational function: $f(x) = \frac{(3x + 8)(x - 1)}{(x - 1)(x + 1)} = \frac{3x + 8}{x + 1}$ for x=-1.

The partition numbers are $-\frac{8}{3}$ and -1. We set up a sign chart with a row for each factor. Store the factor $3x$+8 as Y1 and x+1 as Y2. Store a test number in the calculator and evaluate Y1 and Y2. Write the sign of the result in the chart as shown below. Repeat for the other test numbers.

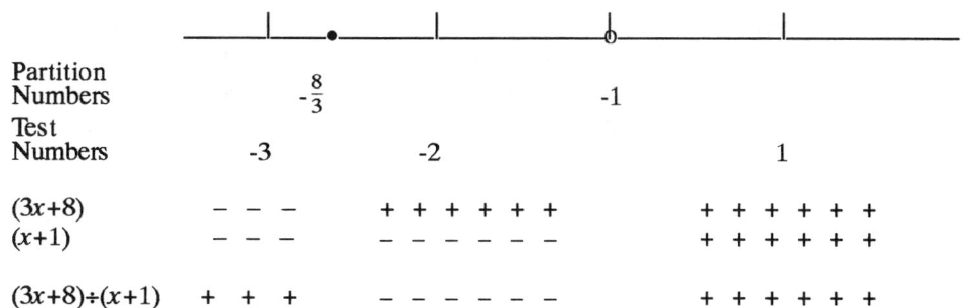

Partition Numbers		$-\frac{8}{3}$		-1	
Test Numbers	-3		-2		1
$(3x+8)$	− − −		+ + + + + +		+ + + + + +
$(x+1)$	− − −		− − − − − −		+ + + + + +
$(3x+8)\div(x+1)$	+ + +		− − − − − −		+ + + + + +

This shows that the function is negative for values of x greater than $-\frac{8}{3}$ and less than -1.

EXAMPLE 2 ***TEXTBOOK SECTION 10-1***

Given the function $f(x) = \frac{|x-2|}{x-2}$. Investigate the behavior of x near 2 to determine whether or not $f(x)$ has a discontinuity at x=2.

Solution: The behavior of $f(x)$ was investigated in Chapter 2 Example 4 and Chapter 9 Example 4 of this manual. The graph is shown to the right. We saw that $\lim_{x\to 2} f(x)$ did not exist. Hence the function is not continuous at x=2.

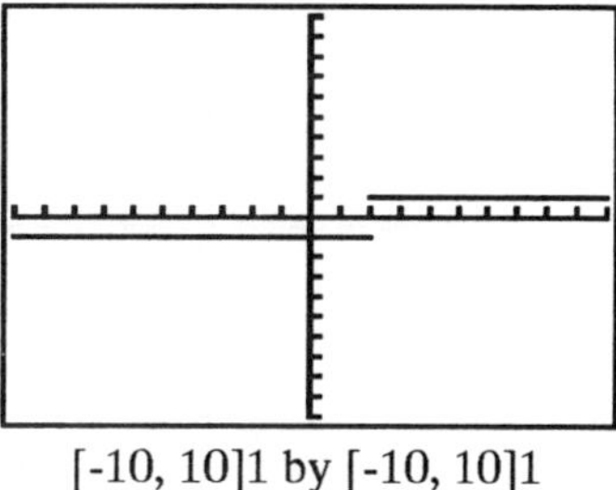

[-10, 10]1 by [-10, 10]1

EXAMPLE 3 ***TEXTBOOK SECTION 10-1***

Given the functions $f(x) = \frac{x^2-4}{x-2}$. Is $f(x)$ continuous at x=2?

Solution: This function was examined in Chapter 2 Example 1 and Chapter 9 Example 3 of this manual. There we found $\lim_{x\to 2} f(x) = 4$ but $f(2)$ is not defined. Hence the function is not continuous at x=2.

EXAMPLE 4 ***TEXTBOOK SECTION 10-1***

Graph f and locate all points of discontinuity.

$$f(x) = \begin{cases} 3.24x - 5 & \text{if } x < 1.8 \\ x^2 & \text{if } x = 1.8 \\ x^3 - 5 & \text{if } x > 1.8 \end{cases}$$

Solution: See Appendix Section A-6, B-6, or C-6 of this manual on how to graph piecewise-defined functions. Graph using [-10, 10]1 by [-10, 10]1. We see a break in the graph near x=1.8 and suspect that the function is discontinuous at this value. Let us investigate this further. Use the cursor and move it along the first piece of the function for values of x close to but less than 1.8. Now move the cursor along the third piece of the function for values of x close to but greater than 1.8. We see the values of the function are close to the same value. Zoom in or set the viewing window variables to [1.75, 1.85].1 by [.7, .9].1 We see the graphs are closer together with just a small break between them. No matter how close we zoom in, the graphs will never meet.

This can be seen by substituting 1.799999 into the first piece of the function and 1.8000001 into the third piece of the function. $f(1.799999) = .83199676$ and $f(1.8000001) = .832$. These values are close together but not equal. No matter how close to 1.8 we choose x the values of the first and third piece of the function will not be equal to $f(1.8) = 1.8^2$. The limit exists and the function is defined at 1.8, but the limit does not equal the function at this value of x. Consequently the function is not continuous at 1.8.

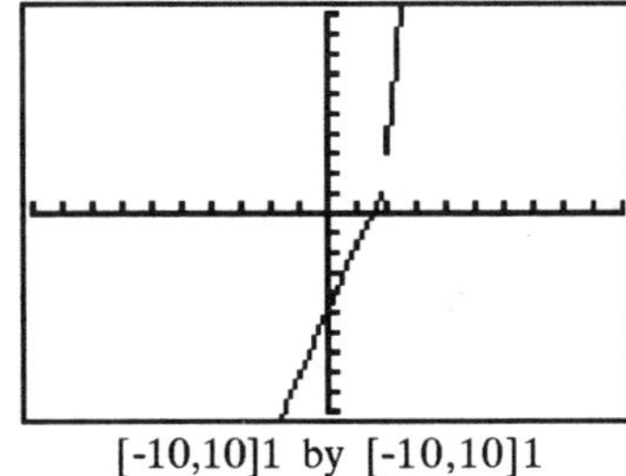
[-10,10]1 by [-10,10]1

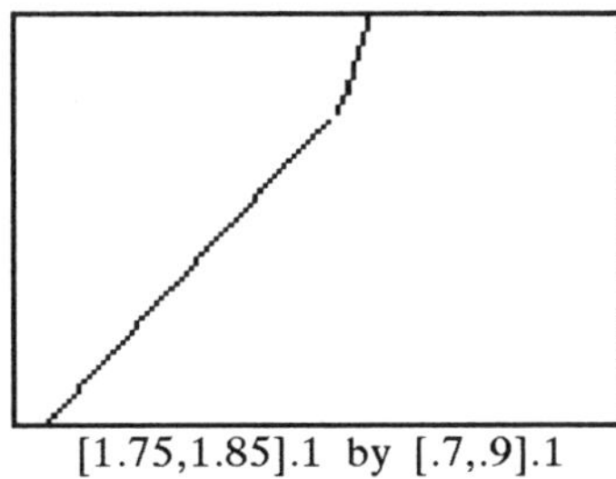
[1.75,1.85].1 by [.7,.9].1

EXAMPLE 5 ***TEXTBOOK SECTION 10-2***

Given the function $f(x) = -2x^3 + 4x^2 - 2.5x - 2$. Use two decimal place accuracy.

(A) Graph using viewing window variables [-10, 10]1 by [-10, 10]1 and [-1, 2]1 by [-3, -2]1.

(B) Fill in the following chart.

x	-1	-.5	0	.5	1.0
$f(x)$					
$f'(x)$					
$f''(x)$					

(C) Find the x intercepts and the y intercepts.

(D) Approximate the local maximum and local minimum points to two decimal places.

(E) Find the intervals over which $f(x)$ is increasing.

(F) Find the intervals over which $f(x)$ is decreasing.

(G) Find any inflection points.

(H) Find the intervals over which $f(x)$ is concave upwards.

(I) Find the intervals over which $f(x)$ is concave downwards.

(J) Show that the concavity changes from positive to negative, or vice versa, when x moves from a little less than the point of concavity to a little greater than the point of concavity.

Solution:

(A) A typical zoom box is shown. This is the region shown in the graph at the right.

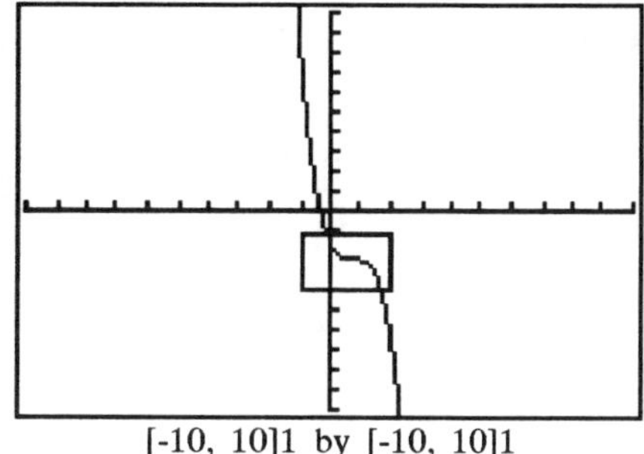

[-10, 10]1 by [-10, 10]1

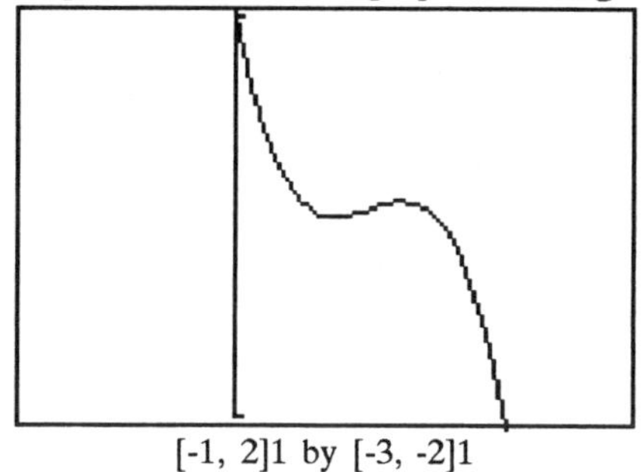

[-1, 2]1 by [-3, -2]1

(B) Store the function in the calculator. Find the first numeric derivative. (See Appendix Section A-15, B-15 or C-15 of this manual.) Some will find the nth derivative. If yours only finds the first numeric derivative, then find the derivative algebraically and store it in the calculator as Y2 and let the calculator find the numeric derivative of this function in order to fill in the bottom line of the chart.

x	-1.0	-0.5	0.0	0.5	1.0
$f(x)$	6.5	.5	-2.0	-2.5	-2.5
$f'(x)$	-16.5	-8.0	-2.5	0.0	-.5
$f''(x)$	20.0	14.0	8.0	2.0	-4.0

(C) Use trace and zoom to find the x intercept to be approximately -.43. Store 0 as X and evaluate the function to find $f(0)$ which is the y intercept. The result is -2.

(D) Method 1 Use trace and zoom to find the local minimum point at (.50, -2.50) and the local maximum point at (.83, -2.46). (See Appendix Section B-15 for the TI-82 to locate the fmin(and fmax(functions.)

Method 2 Graph $f'(x)$ and find where the first derivative is zero by finding the x intercepts of $f'(x)$. To do this, enter $f'(x) = -6x^2 + 8x - 2.5$ as Y2 in the calculator. Deselect Y1. Graph. (See Appendix Section A-6, B-6, or C-6 on how to deselect a function.) $f'(x)=0$ at $x=.50$ and $x=.83$. A local minimum occurs at (.50, -2.50) and a local maximum occurs at (.83, -2.46).

(E) Method 1 Look at the graph of $f(x)$ to determine that the interval over which $f(x)$ is increasing is (.50, .83).

Method 2 Look at the graph of $f'(x)$ to determine where the first derivative is positive. The first derivative is positive when the graph of $f'(x)$ is above the x axis. This is where the function $f(x)$ is increasing.

(F) Method 1 Look at the graph to determine that the intervals over which $f(x)$ is decreasing are $(-\infty, .50)$ and $(.83, \infty)$.

Method 2 Look at the graph of $f'(x)$ to determine where the first derivative is positive or negative. The first derivative is negative when the graph of $f'(x)$ is below the x axis. This is where the function $f(x)$ is increasing.

(G) $f''(x) = -12x + 8$. Graph this function. Use zoom and trace to find the x intercept. It is approximately .67. (The exact value can be found be setting $f'(x)=0$ and solving for x. It is $x=2/3$.) This is the x value at which a point of inflection can occur. Store 0 in the calculator for x and evaluate $f''(x)$. The result is positive. Store 1 in the calculator for x and evaluate $f''(x)$ again. The result is negative. Hence a point of inflection does occur at $x = 2/3$. Store 2/3 as X in the calculator and evaluate $f(x)$. The point of inflection is (.67, -2.48).

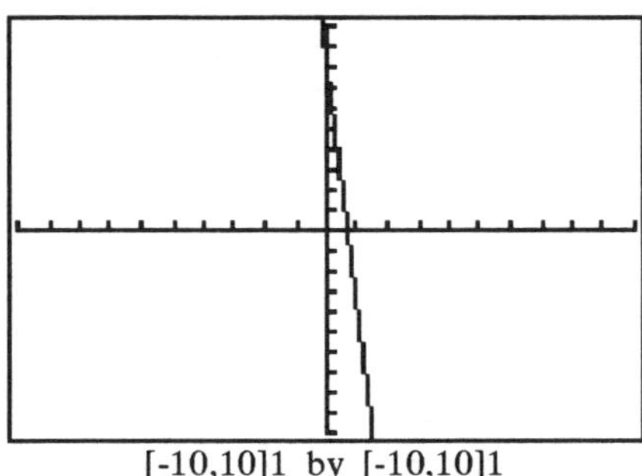
[-10,10]1 by [-10,10]1

(H) & (I) Method 1 Examine the graph of $f''(x)$ to determine where the function is positive. This is where the graph is above the x axis. This occurs when $x < 2/3$. Hence the function is concave upwards on $(-\infty, 2/3)$.

$f''(x) < 0$ for $x > 2/3$ Hence the function is concave downwards on $(2/3, \infty)$.

Method 2 Store $\frac{f''(x)}{|f''(x)|}$ as Y4. [Note: If $f''(x)$ is already stored as Y3, you just have to store Y3÷ ABS (Y3) as Y4.] This gives the sign of the second derivative. The result will be +1 if the sign is positive, and -1 if the sign is negative.

Deselect all functions except the function (stored in Y1) and Y4. Graph. The horizontal line above the x axis shows where $f''(x)$ is positive and $f(x)$ is concave upward. Hence the function is concave upwards on $(-\infty, 2/3)$.

The horizontal line below the x axis shows where $f''(x)$ is negative and $f(x)$ is concave downward. Hence the function is concave downwards on $(2/3, \infty)$.

(J) We see that the graph of $f''(x)$ changes from positive to negative at $x=2/3$. This can be verified by looking at the table: $f''(.5)=2$ and $f''(1) = -4$.

EXAMPLE 6 ***TEXTBOOK SECTION 10-3***

Given $f(x) = x^3$ and $f'(x) = 3x^2$. (A) Graph $f(x)$ using viewing window variables of [0, 3]1 by [0, 6]1. (B) Find an equation of the tangent line to the graph at $x = 1$ and graph it on the same set of coordinate axes. (C) Use the graph to determine if $f(x)$ is concave upward or downward at $x = 1$. (D) Repeat Parts (A)-(C) using $f(x) = (x - 1)^3 + 2$, $f'(x)=3(x-1)^2$, and [0, 3]1 by [0, 6]1.

Solution (A): Enter the function as Y1 as X^3. Set the viewing window variables and graph. The graph is shown to the right.

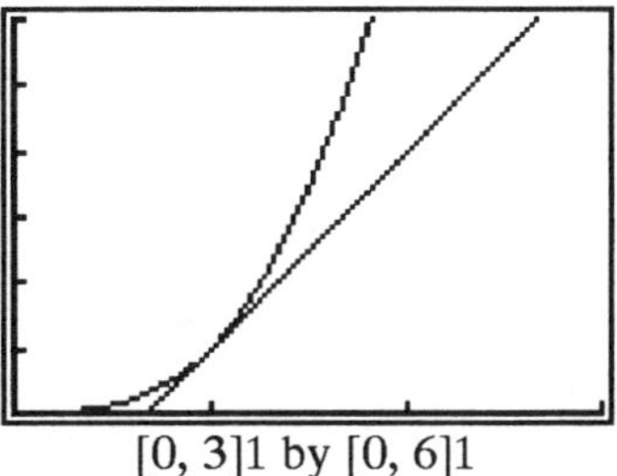
[0, 3]1 by [0, 6]1

Solution (B): The slope of the tangent line is $f'(x) = 3$. The slope-point form of the line is $y-1 = 3(x-1)$. Solving this for y we get $y = 1 + 3(x-1)$. Store this in the calculator as Y2 as 1 + 3 (X - 1) and graph.

Solution (C): Since the graph is entirely above the tangent line, the graph is concave up at $x = 1$. Zoom in or change the viewing window variables to [.5, 1.5].1 by [.5, 1.5].1 to see this.

Solution (D): Enter (X - 1) ^ 3 + 2 as Y1 in the calculator. Set the viewing window variables to [-3, 3]1 by [-6, 6]1. $f'(x)=0$ at $x=1$ and $x=2$. Using the point-slope form of the line we have $y-2=0(x-1)$. So the equation for the tangent line is $y = 2$. The graph is shown to the right.

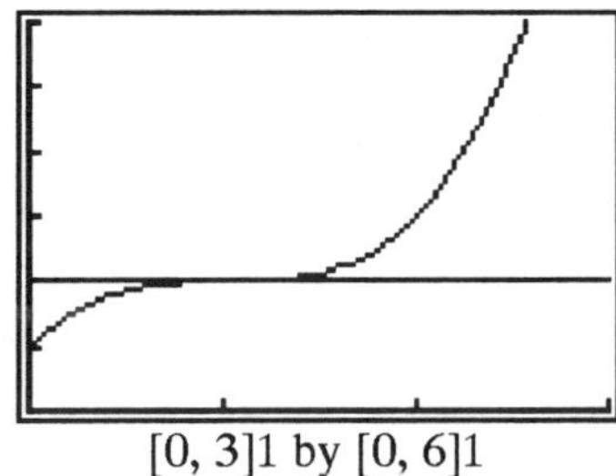
[0, 3]1 by [0, 6]1

Here we see the graph of the function to be below the tangent line for $x < 1$ and above the tangent line for $x > 1$. Hence, a point of inflection occurs at $x = 1$.

EXAMPLE 7 ***TEXTBOOK SECTION 10-4***

Use graphing techniques to find $\lim_{x \to \infty} \dfrac{3x^2 - x + 2}{5x^2 - 8}$.

Solution: Store the expression in the calculator as (3 X ^ 2 - X + 2) ÷ (5 X ^ 2 - 8). Since we are interested in very large values of x graph using [10,100]5 by [-10,10]1. Use trace to see what the values y are for large values of x. We see $f(x)$ has values about .6. Change the viewing window variables to [10, 300]0 by[.5, .7].1 to get a better look. We now see the graph rising slightly as x increases. Using trace we see that the y values get close to .6 as x gets larger and larger.

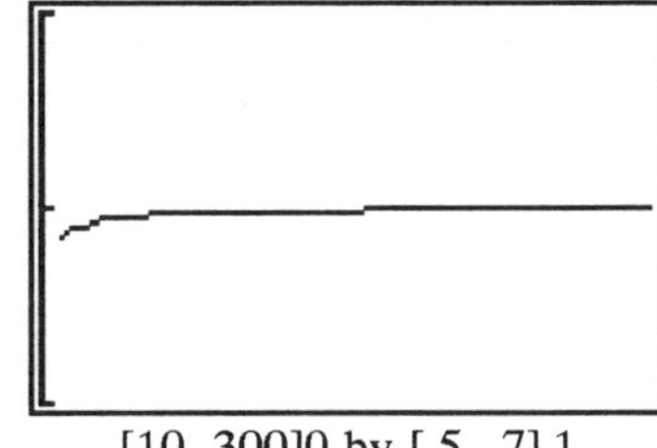
[10, 300]0 by [.5, .7].1

Hence, $\lim_{x \to \infty} \dfrac{3x^2 - x + 2}{5x^2 - 8} = .6$.

EXAMPLE 8

Graph $f(x) = 20 \log_3(x + e^{.02x}) - .06x + 2$. Find all local maxima, local minima, x intercepts, y intercepts, and asymptotes by examining the graph.

Solution: The change-of-base formula must be used in order to graph this function. It is: $\log_b x = \frac{\ln x}{\ln b}$. Hence the function can be written as $f(x) = 20 \frac{\ln (x + e^{.02x})}{\ln 3} - .06x + 2$. Enter the function in the calculator as:

20 ([LN] (X + [e^] (.02 X))) ÷ [LN] 3 - .06 X + 2

where the [LN] means the built-in natural logarithm function in the calculator. Note that [e^] is built in your calculator above the [LN] key.

Graph using [-10, 10]1 by [-10, 10]1. We see that different limits on the x and y axes would give a better display of this function. Evaluating this function at a few values of x may help: $f(-5)$ is not defined (an error message on the calculator); $f(-1)$ is not defined; $f(-.95)=-22.0$; $f(-.5)=27.6$; $f(0)=40$; $f(1)=51.6$; $f(10)=72.0$; and $f(50)=52.2$. So we see that a possible starting viewing window could be [-2, 20]2 by [-50, 100]5. It appears that there is no local minima or local maxima.

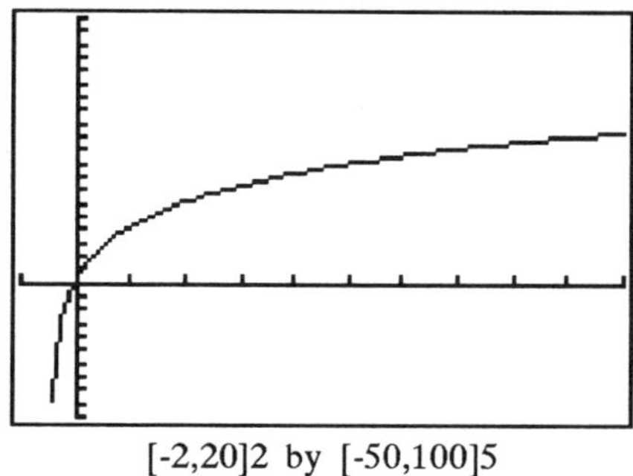

[-2,20]2 by [-50,100]5

By using trace and zoom (or adjusting the viewing window variables) we can find the x intercept. It is -.10. Evaluating the function at $x = 0$ shows the y intercept is 2.

To investigate the existence of a horizontal asymptote set the viewing window variables to [-2, 20]2 by [-50, 100]5 and graph the function. It appears that the graph may be approaching a horizontal asymptote. Setting the viewing window variables to [50, 100]10 by [0, 100]10 indicates that $f(x)$ is still increasing. Set the viewing window variables to [500, 2000]100 by [0, 1000]30 to investigate this further. There is no horizontal asymptote.

Use trace and zoom in on the left end of the graph. After several zooms, or setting the viewing window variables at [-.98057948, -.98057945].00000001 by [-500,-300]10, you will notice that the x value gets close to -.9805795 and the y value continues to get large negatively. This indicates that there is a vertical asymptote at $x \approx -.9805795$.

Exercise Set 10

1. Given the functions in Problems 27-36 of Textbook Exercise 10-1. (A) Graph using the viewing rectangle variables as described in Example 1(C) above. (B) Describe the behavior at each point of discontinuity. (C) Identify all vertical asymptotes. (D) Write the right-hand and left-hand limits using limit notation for each point of discontinuity.

2. Solve Problems 37-42 of Textbook Exercise 10-1 using graphing techniques and a sign chart as described in Example 2 above.

3. Solve Problems 43-48 of Textbook Exercise 10-1.

4. Solve Problems 49-56 of Textbook Exercise 10-1 by graphing using viewing window variables of [-10, 10]1 by [-10, 10]1 and zooming as necessary.

5. Solve Problems 57-62 of Textbook Exercise 10-1 by graphing. Write the right-hand and left-hand limits using limit notation for each point of discontinuity.

6. Graph the greatest integer function described in Problems 65-66 of Textbook Exercise by entering [IPart] X as the function and using the viewing window variables described in Example 1(C) above. Solve the problems as stated in the textbook.

7. Is $f(x) = \dfrac{2x^2 - 7x + 5}{x^2 - 2x + 1}$ continuous at $x = -1$?

8. Is $f(x) = \dfrac{2x^2 - 3x - 5}{x^2 - 4x + 3}$ continuous at $x = 1$? at $x = 3$.

9. Is $f(x) = \dfrac{x - 2}{|x - 2|}$ continuous at $x = 2$?
   (Compare to Example 2 above. Is this the same function?)

10. Is $f(x) = \dfrac{|x + 3|}{x + 3}$ continuous at $x = -3$?

11. Graph f and locate all points of discontinuity.

$$f(x) = \begin{cases} 4.2x + 2 & \text{if } x < -.5 \\ x^2 + 3 & \text{if } x = -.5 \\ x^3 - 2 & \text{if } x > -.5 \end{cases}$$

12. Graph f and locate all points of discontinuity.

$$f(x) = \begin{cases} x - 4 & \text{if } x < 3 \\ x^2 - 3 & \text{if } x \geq 3 \end{cases}$$

13. Solve Problems 25-38 of Textbook Exercise 10-2 by graphing and using zoom and trace, or other calculator methods, to find the local extrema.

14. Solve Problems 39-44 of Textbook Exercise 10-2 by graphing and using zoom and trace, or other calculator methods, to find the local extrema.

15. Solve Problems 45-48 of Textbook Exercise 10-2.

16. Solve Problems 57-70 of Textbook Exercise 10-2 using graphing techniques.

17. In Problems 49-56 of Textbook Exercise 10-3 graph the function and the tangent line at the point of inflection.

18. In Problems 49-56 of Textbook Exercise 10-3 find an equation for and graph the tangent line at $x = 2$ for each of the functions. Use the graph to determine if the function is concave upward or concave downward at $x = 2$.

19. Solve Problems 69-72 of Textbook Exercise 10-3.

20. Solve Problems 77-80 of Textbook Exercise 10-3 by graphing.

21. Solve Problems 25-34 of Textbook Exercise 10-4 by graphing.

22. Solve Problems 35-48 of Textbook Exercise 10-4 by graphing.

23. Solve Problems 49-60 of Textbook Exercise 10-4 by graphing.

24. Solve Problems 65-78 of Textbook Exercise 10-4 by graphing.

25. Solve Problems 79-84 of Textbook Exercise 10-4 by graphing.

26. In Problems 79-84 graph the function and the tangent line at $x = 0$. Use this information to determine if the function is concave upward or concave downward at $x = 0$. Check your answer by evaluating the second derivative at $x = 0$.

27. Given the function $f(x) = x^3 - 2.5x^2 + 2x + 1$. Use two decimal place accuracy.

(A) Graph using [-10, 10]1 by [-10, 10]1 and [-1, 2]1 by [0, 2]1.
(B) Fill in the following chart.

x	0	.4	.6	.8	1.0	1.2
$f(x)$						
$f'(x)$						
$f''(x)$						

(C) Find the x intercepts and the y intercepts.
(D) Approximate the maximum and minimum points to two decimal places.
(E) Find the intervals over which $f(x)$ is increasing.
(F) Find the intervals over which $f(x)$ is decreasing.
(G) Find any inflection points.
(H) Find the intervals over which $f(x)$ is concave upwards.
(I) Find the intervals over which $f(x)$ is concave downwards.
(J) Show that the concavity changes from positive to negative, or vice versa, when x moves from a little less than the point of concavity to a little greater than the point of concavity.

28. Given the function $f(x) = x^4 - 3x^3 + 2x^2$. Use two decimal place accuracy.

(A) Graph using [-10, 10]1 by [-10, 10]1 and [-1, 2]1 by [-1, 1]1.
(B) Fill in the following chart.

x	-1.0	-0.5	0.0	0.2	0.6	1.0
$f(x)$						
$f'(x)$						
$f''(x)$						

(C) Find the x intercepts and the y intercepts.
(D) Approximate the maximum and minimum points to two decimal places.
(E) Find the intervals over which $f(x)$ is increasing.
(F) Find the intervals over which $f(x)$ is decreasing.

(G) Find any inflection points.
(H) Find the intervals over which $f(x)$ is concave upwards.
(I) Find the intervals over which $f(x)$ is concave downwards.
(J) Show that the concavity changes from positive to negative, or vice versa, when x moves from a little less than the point of concavity to a little greater than the point of concavity.

29. Use graphing techniques to find any horizontal asymptotes of $f(x) = \dfrac{4x^3 - 2x + 3}{7x^2 + 3}$.

30. Use graphing techniques to find any horizontal asymptotes of $f(x) = \dfrac{7x^2 + x - 5}{5x^4 - 2}$.

31. Find any asymptotes of $f(x) = \dfrac{2x^2 + 3x - 4}{x^2 - 5}$.

32. Find any asymptotes of $f(x) = \dfrac{5x^2 + 3x - 3}{x^2 + 1}$.

33. Graph $f(x) = 1.5 \log_2(x - e^{.03x}) + .05x + 2$. Find all maxima, minima, x intercepts, y intercepts, and asymptotes by examining the graph.

34. Graph $f(x) = e^{.01x + \ln x} + .03x + 2$. Find all maxima, minima, x intercepts, y intercepts, and asymptotes by examining the graph.

35. Solve Problems 9-22 of Textbook Exercise 10-5.

36. Do Textbook Chapter 10 Group Activity. Use the calculator where appropriate.

Chapter 11

Additional Derivative Topics

This chapter contains examples using the calculator to illustrate :

- Exploring the definition of e
- Comparing compound interest and continuously compound interest
- Solving differential equations
- Exploring the relationship between $y = e^x$ and $y = \ln x$
- Comparing the graphs logarithm functions to different bases
- Exploring the derivative of the exponential function

The graphing calculator features used are:

- Calculation
- Replay key
- Storing a function
- Evaluating a function
- Graphing and viewing window variables
- Tracing and zooming

EXAMPLE 1 ***TEXTBOOK SECTION 11-1***

Graph the function $f(x) = (1 + x)^{\frac{1}{x}}$ for $-0.5 < x < .5$. Use trace and zoom to find, to three decimal places, $\lim_{x \to 0} (1 + x)^{\frac{1}{x}}$.

Solution: Enter the expression in the calculator as Y1 as (1 + X) ^ (1 ÷ X) . Set the viewing rectangle variables to [-.5, .5].1 by [0, 4]1. Graph. Use trace and zoom to see that the values of $f(x)$ are near 2.718 for x near 0.

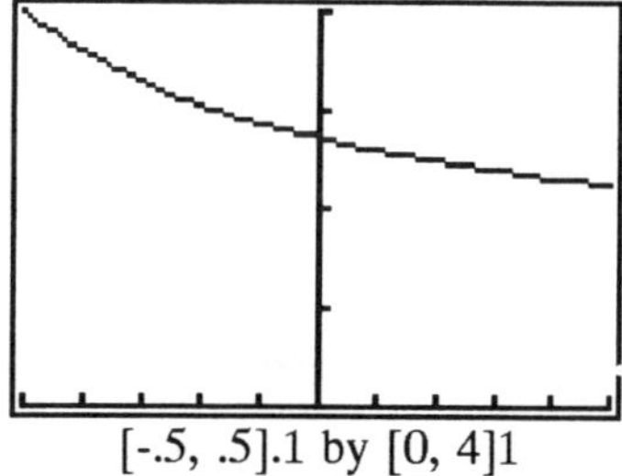

[-.5, .5].1 by [0, 4]1

EXAMPLE 2 **TEXTBOOK SECTION 11-1**

$10000 is deposited in an account.

(A) Calculate the amount in the account at the end of 30 years if the account earns 8% simple interest.

(B) Calculate the amount in the account at the end of 30 years if the account earns 8% interest compounded yearly.

(C) Calculate the amount in the account at the end of 30 years if the account earns 8% interest compounded monthly.

(D) Calculate the amount in the account at the end of 30 years if the account earns 8% interest compounded daily.

(E) Calculate the amount in the account at the end of 30 years if the account earns 8% interest compounded continuously.

(F) Graph the growth of the accounts for Part (B), (C), (D), and (E) using viewing window variables of [0, 30]5 by [0, 111000]0. Compare the graphs. To see the difference between the graphs for daily and continuous compounding set the viewing window variables at [29.9, 30].1 by [109000, 110400]0.

Solution (A): The formula for calculating future value for simple interest is $A = P(1+rt)$ where P=principal, r=yearly interest rate, and t=number of years. For this situation we have $10000(1+.08\times 30) = 34000$. $34,000 is in the account after 30 years.

Solution (B): The formula for calculating the future value for compound interest is $A = P(1+\frac{r}{n})^{nt}$ where P=principal, r=yearly interest rate, n=number of compounding periods in one year, and t=number of years. For this situation we have

$10000\left(1+\frac{.08}{1}\right)^{1\times 30} = 100626.57$. There is $100,626.57 in the account after 30 years.

Solution (C): Using the formula in Part (B) we have $10000\left(1+\frac{.08}{12}\right)^{12\times 30} =$ 109357.30. There is $109,357.30 in the account after 30 years.

Solution (D): Using the formula in Part (B) we have $10000\left(1+\frac{.08}{365}\right)^{365\times 30} =$ 110202.78. There is $110,202.78 in the account after 30 years.

Solution (E): The formula for continuous compounding is $A = Pe^{rt}$. In this situation we have $10000e^{.08\times30} = 110231.76$. There is $110,231.76 in the account after 30 years.

Solution (F): Store the following in the function list in the calculator as:

Y1	10000 (1 + .08 ÷ 1) ^ (1 X)
Y2	10000 (1 + .08 ÷ 12) ^ (12 X)
Y3	10000 (1 + .08 ÷ 365) ^ (365 X)
Y4	10000 [2nd] [LN] (.08 X)

Set the viewing window variables to [0, 30]5 by [0, 111000]0 and graph. Watch very closely as Y3 and Y4 are graphed. Since the range is so wide for the y variable we see very little difference between these curves.

Change the viewing window variables to [29.9, 30].1 by [109000, 110400]0 to see the difference between the graphs for the last tenth of the thirtieth year. We see there is little difference between daily and continuously compounded interest.

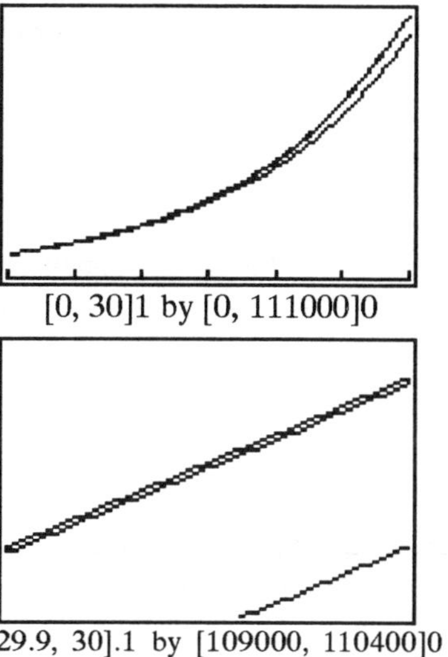

[0, 30]1 by [0, 111000]0

[29.9, 30].1 by [109000, 110400]0

EXAMPLE 3 — *TEXTBOOK SECTION 11-2*

Given $f(x) = e^x$.

(A) Graph $y = e^x$, its inverse, and $y = x$ on the same set of coordinate axes graph using [-4.7, 4.8]1 by [-1, 5.3]1 on the TI-81, [-4.7, 4.7]1 by [-1, 5.2]1 on the TI-82, or [-6.3, 6.3]1 by [-1, 5.2]1 on the TI-85. .

(B) Calculate, to two decimal places, the y coordinate for $y = e^x$ when $x = 1$.

(C) Find the coordinates, to two decimal places, of the inverse of $y = e^x$ that correspond with those found in Part (B).

(D) Discuss the relationship between these two pairs of coordinates found in Parts (B) and (C) above. (See Chapter 2 Example 8 of this manual where we graphed $y = 3^x$, its inverse, and $y = x$ on the same set of coordinate axes.)

(E) Find, to two decimal places, the slope of the tangent line at each of the points found in Parts (B) and (C) above.

Solution (A): Enter the function in the calculator as Y1 as [2nd] [LN] X. Enter the inverse function, $y = \ln x$, as Y2 as [LN] X. Enter the function $y = x$ as Y3 as X.

Solution (B): Calculate this by entering [e^] 1 [ENTER] on the home screen. The coordinates of the point are (1, 2.72). These points are shown as large dots in the graph below. However, you will not be able to plot them as such on your calculator.

Solution (C): The coordinates of the point on the graph of the inverse of $y = e^x$ are the same as those for the function $y = e^x$ but in reverse order. They are (2.72, 1). This can be checked by finding (2.72, ln 2.72).

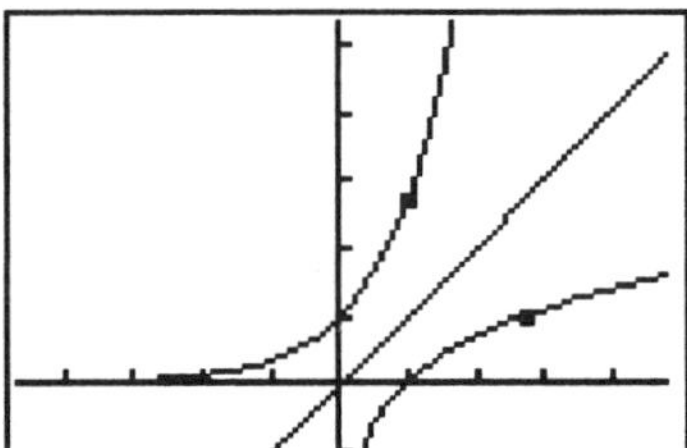

Solution (D): The coordinates are in reverse order of each other because of the relationship of a function with its inverse.

Solution (E): The slope of the tangent line is given by the first derivative. For $f(x) = e^x$ we have $f'(x) = e^x$. So the slope of the tangent line is $e^1 = 2.72$.

For $y = \ln x$ we have $y' = \frac{1}{x}$. So the slope of this tangent line is $\frac{1}{1} = 1$.

EXAMPLE 4 ***TEXTBOOK SECTION 11-2***

Given the function $f(x) = e^x$.

(A) Graph using [-4.7, 4.8]1 by [-1, 5.3]1 on the TI-81, [-4.7, 4.7]1 by [-1, 5.2]1 on the TI-82, or [-6.3, 6.3]1 by [-1, 5.2]1 on the TI-85.

(B) Draw a tangent line with one end of the line segment at (2, 3).

(C) Estimate the slope of the tangent line.

(D) Estimate, to one decimal place, the coordinates of the point of tangency.

(E) Check your estimate of the slope by evaluating $f(x)$ at x value of the point of tangency.

Solution (A): Enter the function in the calculator as Y1 as e^ X. Set the viewing window variables and graph.

Solution (B): While the graph screen is displayed, get Line(option from the DRAW menu on the TI-81 or TI-82. On the TI-85 press [MORE] [F2] <DRAW> [F2] <LINE>.

Use the arrow keys to move the cursor to (2, 3) and press [ENTER] to anchor the right-hand endpoint of the line segment. Now use the arrow keys again to position the left-hand endpoint of the line segment so that it looks like it is tangent to the curve. One such point may be (-.8, 0). Press [ENTER] again to anchor the left-hand endpoint.

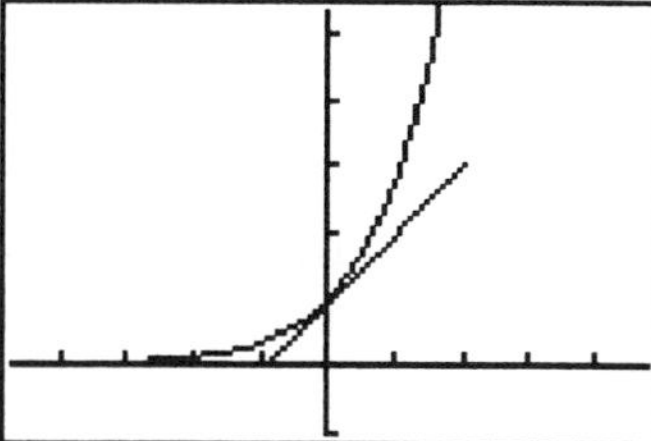

Solution (C): The slope of this line segment is $m = \frac{3 - 0}{2 - (-.8)} = 1.1$.

Solution (D): Use trace to estimate the point of tangency to be approximately (.2, 1.2).

Solution (E): $f'(x) = e^x$. So, $f'(x) = e^{.2} = 1.22$.

EXAMPLE 5 ***TEXTBOOK SECTION 11-3***

(A) Graph $y = \log_2 x$, $y = \ln x$, and $y = \log_3 x$ on the same set of coordinate axes using [-.7, 8.8]1 by [-4, 2.3]1 on the TI-81, [-.7, 8.7]1 by [-4, 1.2]1 on the TI-82, or [-2.3, 10.3]1 by [-4, 1.2]1 on the TI-85.

(B) Describe the relationship of these graphs for $x > 1$.

(C) Describe the relationship of these graphs for $0 < x < 1$.

(D) Describe the relationship of these graphs at $x = 1$.

Solution (A): We need to use the identity $\log_b x = \frac{\ln x}{\ln b}$. So $\log_2 x = \frac{\ln x}{\ln 2}$ and $\log_3 x = \frac{\ln x}{\ln 3}$. Store the functions as:

Y1	([LN] X) ÷ ([LN] 2)
Y2	([LN] X)
Y3	([LN] X) ÷ ([LN] 3)

Graph.

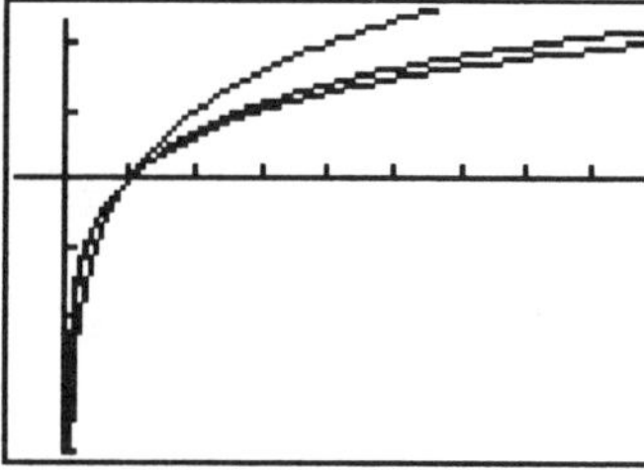

Solution (B): The graph of $y = \log_3 x$ is above the graph of $y = \ln x$ which is above the graph of $y = \log_2 x$.

Solution (C): The graph of $y = \log_2 x$ is above the graph of $y = \ln x$ which is above the graph of $y = \log_3 x$. This is a little hard to see. Change the viewing window variables to [0, 1].1 by [-5, 0]1 to see this better.

Solution (D): All of the graphs pass through the point (1, 0).

Solution (E): The slope of the tangent line for $y = \log_3 x$ is greater than for $y = \ln x$ which will be greater than $y = \log_2 x$. This can be checked out by finding the numerical derivative on the calculator for each of these functions at a point, say $x = 3$. The slopes are .48, .33, and .30, respectively.

Exercise Set 11

1. Solve Problems 1-2 of Textbook Exercise 11-1.
2. Solve Problems 3-10 of Textbook Exercise 11-1.
3. Solve Problems 13-24 of Textbook Exercise 11-1.
4. Solve Problems 25-28 of Textbook Exercise 11-1.
5. Solve Problems 39-52 of Textbook Exercise 11-2 by graphing.
6. Solve Problems 53-58 of Textbook Exercise 11-2 by graphing.
7. Solve Problems 47-52 of Textbook Exercise 11-3.
8. Solve Problems 67-68 of Textbook Exercise 11-3.
9. Solve Problems 71-72 of Textbook Exercise 11-3.
10. Do Textbook Chapter 11 Group Activity. Use the calculator where appropriate.

Chapter 12

Integration

This chapter contains examples using the calculator to illustrate :

- Graphing antiderivatives of a function
- Calculating continuously compound interest
- Evaluating definite integrals
- Finding the area between two intersecting curves
- Using the rectangle rule

The graphing calculator features used are:

- Calculation
- Replay key
- Storing a function
- Evaluating a function
- Graphing and viewing window variables
- Tracing and zooming

EXAMPLE 1 ***TEXTBOOK SECTION 12-1***

(A) Graph $y = x^2+1$, and $y = x^2-1$ viewing window variables at [-1.35, 3.4]1 by [-2, 4.3]1 for the TI-81, [-1.3, 3.4]1 by [-2, 4.2]1 for the TI-82, or [-2.9, 3.4]1 by [-2, 4.2]1 for the TI-85.

(B) Draw the tangent line to each graph at $x = 1$.

(C) Explain why the slope of the tangent line is the same in each case.

Solution (A): Enter the functions into the calculator as:

Y1 X ^ 2 + 1
Y2 X ^ 2 - 1

Set the viewing window variables and graph.

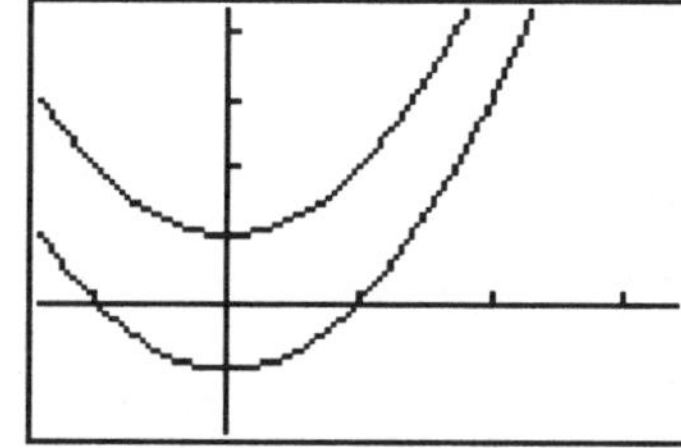

Solution (B): The slope of the tangent line is given by $y' = 2x$. The slope of the tangent line is 2 at x=1. The equations of the tangent lines and two points on the tangent line are:

Function	Point	Equation of the tangent line	Two points on the tangent line
$y = x^2+1$	(1, 2)	$y - 2 = 2(x - 1)$	(0, 0) and (2, 4)
$y = x^2 - 1$	(1, 0)	$y + 0 = 2(x - 1)$	(0, -2) and (2, 2)

While the graph screen is displayed, get the Line(option. This is found on the DRAW menu on the TI-81 and TI-82. On the TI-85 press [MORE] [F2] <DRAW> [F2] <LINE>. (See Appendix Section A-3, B-3, or C-3 of this manual.)

Use the arrow keys to move the cursor to (0, 0) and press [ENTER] to anchor the left-hand endpoint of the line segment. Now use the arrow keys again to position the right-hand endpoint of the line segment at (2,4) and press [ENTER] again to anchor the left-hand endpoint. Repeat for the other line segment. The result is shown at the right.

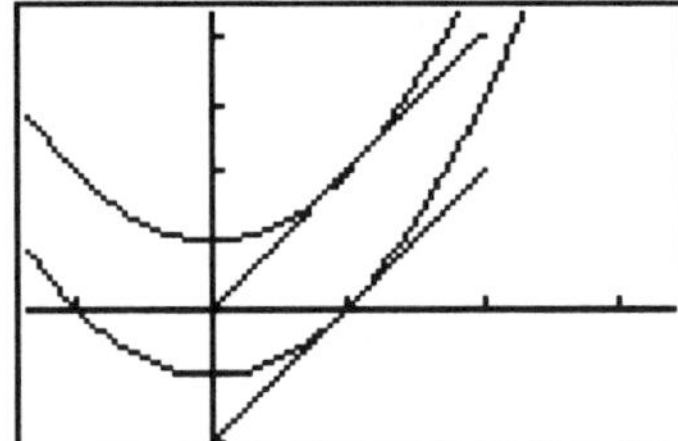

Another method for drawing the tangent lines in this case would be to store the equations as Y3 as 2 (X - 1) + 2 and as Y4 as 2 (X - 1) and graph.

Solution (C): The slope of the tangent line is the same for both of these functions because the derivative is the same for both of the functions.

EXAMPLE 2 — *TEXTBOOK SECTION 12-1*

Given $f'(x) = \sqrt{x - 1} - 2.5$:

(A) Graph $f'(x)$ using [0, 10]1 by [-5, 5]1.

(B) Use trace and zoom, or other calculator methods, to determine over what intervals $f(x)$ is increasing? Use two decimal place accuracy.

(C) Use trace and zoom, or other calculator methods, to determine over what intervals $f(x)$ is decreasing? Use two decimal place accuracy.

(D) Find the critical values. Use two decimal place accuracy.

(E) Determine if a maximum or minimum occurs at each of the critical values.

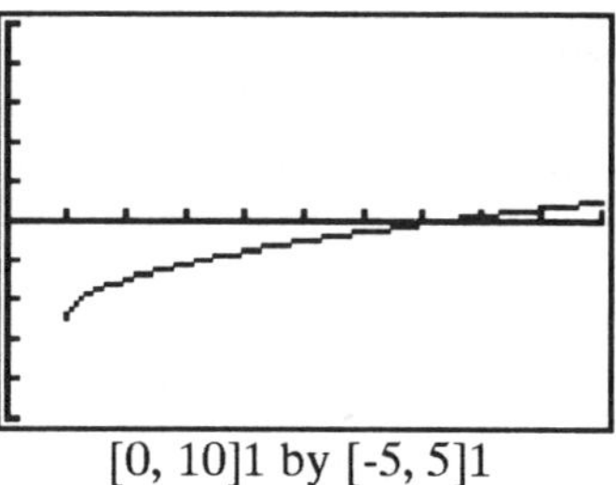
[0, 10]1 by [-5, 5]1

Solution (A): Enter $f'(x)$ as Y1 as 2nd $\sqrt{\ }$ (X - 1) - 2.5 . Set the viewing window variables at [0, 10]1 by [-5, 5]1. Graph.

Solution (B): We need to determine the partition numbers. We can use zoom and trace to find the x value of the left endpoint of the graph and the x intercept. The left endpoint occurs at $x=1$. The x intercept is x = 7.5. Hence, the partition numbers are 1 and 7.25.

An alternative method would be to algebraically determine the domain of $f'(x)$ as D = $\{x \mid x \geq 1\}$. This establishes the left partition number. It is $x = 1$.

Solving $\sqrt{x - 1} - 2.5 = 0$ will determine the other partition number. This can be done algebraically or on the calculator by finding the x intercept of this function. It is 7.25.

We see that the graph of $f'(x)$ is positive for $x > 7.25$. Hence the function $f(x)$ is increasing for $x > 7.25$.

Solution (C): We see that the graph of $f'(x)$ is negative for $1 < x < 7.25$. Hence the function $f(x)$ is decreasing for $1 < x < 7.25$.

Solution (D): The critical value is the x intercept of $f'(x)$. The critical value is 7.25.

Solution (E): A local minimum occurs at the critical value since the function is decreasing to the left of x=7.25 and is increasing to the right of x=7.25.

EXAMPLE 3 — *TEXTBOOK SECTION 12-1*

Given the derivative $f'(x) = x^3$:

(A) Graph the derivative using [-5, 5]1 by [-5, 5]1.

(B) Graph the second derivative using the built-in function nDeriv(with X in place of the value as Y2.

(C) Graph the anti-derivative with the constant of integration equal to 0.

(D) Trace along the second derivative curve to determine the points of inflection, over which intervals the function is concave upward, and over which intervals the function is concave downward.

(E) Trace along the derivative curve and compare the value of $f'(x)$ to the slope of the graph of $f(x)$.

(F) Graph the anti-derivative $f'(x)$ satisfying $f(0) = 3$.

Solution (A): Set the viewing window variables at the specified values. Enter the $f'(x)$ in the function list as Y1 as X ^ 3. Graph. See the graph to the right.

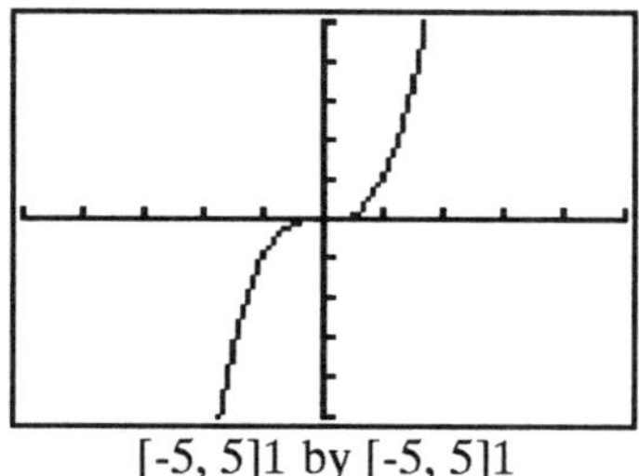

[-5, 5]1 by [-5, 5]1

Solution (B):

TI-81 Enter the second derivative function as Y2 as

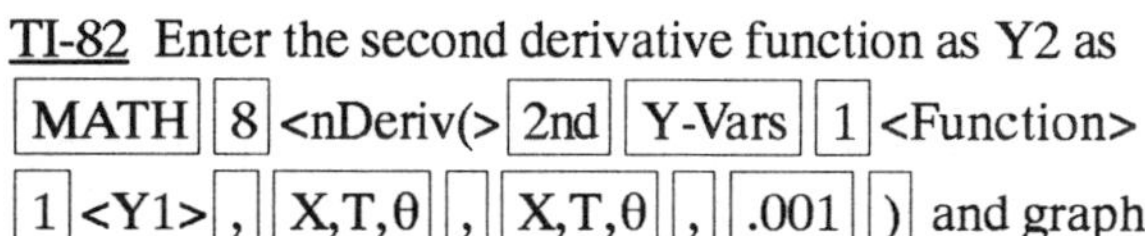

MATH 8 <NDeriv(> 2nd Y-Vars 1 <Y1> ALPHA , X|T) and graph.

TI-82 Enter the second derivative function as Y2 as

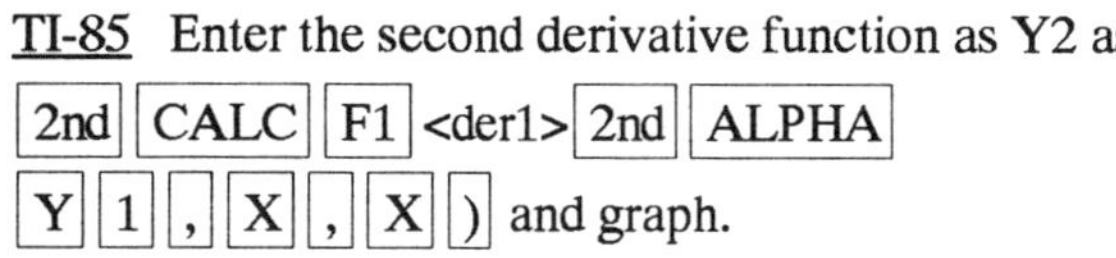

MATH 8 <nDeriv(> 2nd Y-Vars 1 <Function> 1 <Y1> , X,T,θ , X,T,θ , .001) and graph.

TI-85 Enter the second derivative function as Y2 as 2nd CALC F1 <der1> 2nd ALPHA Y 1 , X , X) and graph.

The derivative of the function stored as Y1 will be graphed. See the graph to the right.

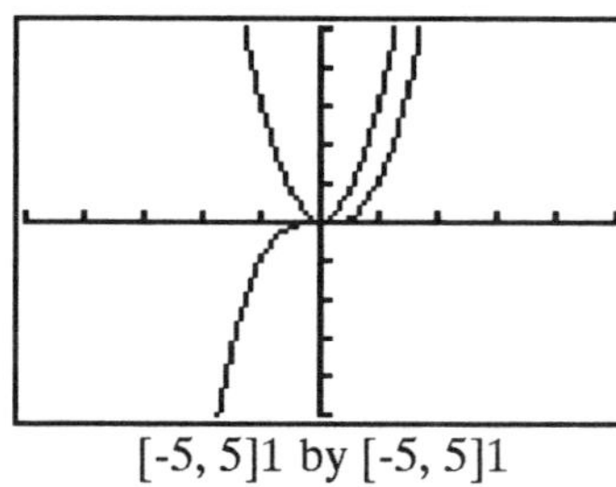

[-5, 5]1 by [-5, 5]1

Solution (C):

TI-81 Integration cannot be performed on the TI-81. The anti-derivative can be found algebraically, entered as Y3, and graphed. The anti-derivative is $f(x) = \frac{x^4}{4}$. Enter this as Y3 as (X|T ^ 4) ÷ 4 .

TI-82 Enter the anti-derivative function with constant zero as Y3 as MATH 9 <fnInt(>

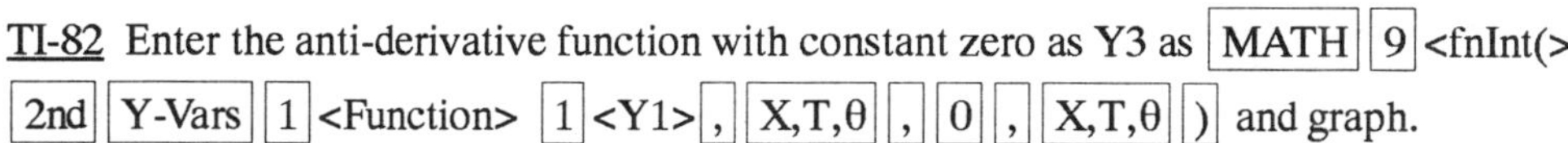

2nd Y-Vars 1 <Function> 1 <Y1> , X,T,θ , 0 , X,T,θ) and graph.

TI-85 Enter the anti-derivative function with constant zero as Y3 as 2nd CALC F5 <fnInt> 2nd ALPHA Y 1 , x-VAR , 0 , x-VAR) and graph. The graph is shown at the right.

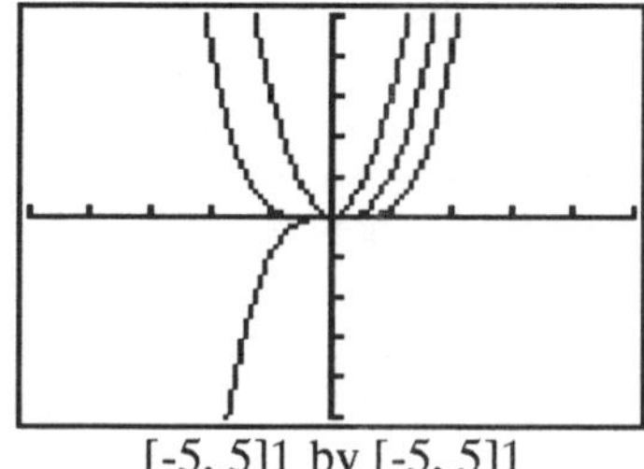

[-5, 5]1 by [-5, 5]1

Solution (D): There is no point of inflection since the second derivative curve graphed in Part (B) above is always positive. This means that the function $f(x)$ is always concave upward.

Solution (E): As we trace along the derivative curve ($f'(x)$ stored as Y1), we see the values are negative for $x < 0$ and positive for $x > 0$. This tells us that the function is decreasing for $x < 0$ and that the graph of $f(x)$ is falling for $x < 0$. Similarly, $f'(x)$ is positive for $x > 0$ which means that $f(x)$ is increasing and the graph of $f(x)$ is rising for $x > 0$.

Solution (F): (0, 3) is the y intercept of $f(x)$. Hence we just need to add 3 to the expression in Part (C) above.

<u>TI-81</u> Store $\frac{x^4}{4}$ + 3 as Y4 as (X|T ^ 4) ÷ 4 + 3 and graph.

<u>TI-82</u> Store the function as Y4 as MATH 9 <fnInt(> 2nd Y-Vars 1 <Function> 1 <Y1> , X,T,θ , 0 , X,T,θ) + 3 and graph.

<u>TI-85</u> Enter the anti-derivative function with constant 3 as Y4 as 2nd CALC F5 <fnInt> 2nd ALPHA Y 1 , x-VAR , 0 , x-VAR) + 3 and graph. The graph is shown at the right.

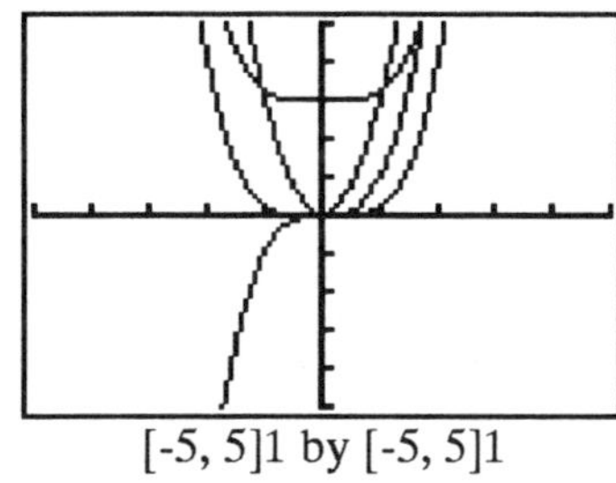

[-5, 5]1 by [-5, 5]1

EXAMPLE 4 ***TEXTBOOK SECTION 12-3***

A piece of human bone was found at an archaeological site in Africa. If 10% of the original amount of radioactive carbon-14 was present, estimate the age of the bone to the nearest 100 years.

Solution: Using the exponential growth law for $\frac{dQ}{dt} = 0.000\ 123\ 8Q$ and $Q(0) = Q_0$ we find that $Q = Q_0\ e^{-0.0001238t}$. Our problem is to find t so that $Q = 0.1Q_0$.

Substituting, we get

$$.01Q_0 = Q_0 e^{-0.0001238t}$$

$$0.1 = e^{-0.0001238t}$$

To solve this equation graphically, store the function $0.1 - e^{-0.0001238t}$ in the calculator as 0.1 - e^ (⁻.0001238 X). Graph using [0,30000]1000 by [-.5,.5].1. The viewing window variables were chosen by first graphing the expression for several different viewing window variables and choosing one that showed the x intercept clearly. Use trace and zoom to find the x intercept. This value is 18600. Hence, the solution to the problem t = 18,600 years.

[0,30000]1000 by [-.5,.5].1

EXAMPLE 5 ***TEXTBOOK SECTION 12-5***

Evaluate $\int_{1.38}^{2.96} 4x - 2e^x + \frac{5}{x}\, dx$. Use two decimal place accuracy.

Solution: Some graphing calculators have built-in features to do numeric integration. (See Appendix Section B-15 and C-15 of this manual.) However, this example uses the method of evaluating an antiderivative of the integrand. An antiderivative is $\frac{4x^2}{2} - 2e^x + 5 \ln x$. Store the antiderivative as a function in the calculator as (4 X ^ 2) ÷ 2 - 2 e^ X + 5 [LN] X . Store 2.96 as X and evaluate the expression. The result is -15.65. Store the result in a memory location, say A. Store 1.38 as X and evaluate the expression again. The result is -2.53. Calculate A - Ans. The result is -13.12.

$$\int_{1.38}^{2.96} 4x - 2e^x + \frac{5}{x}\, dx = \left. \frac{4x^2}{2} - 2e^x + 5 \ln x \right|_{1.38}^{2.96} = -15.65 - (-2.53) = -13.12.$$

EXAMPLE 6 ***TEXTBOOK SECTION 12-5***

Approximate the integral below, to two decimal places, using the midpoint sum and n=4.

$$\int_0^2 \frac{4x}{x+6}\, dx$$

Solution: We find Δx by calculating $\frac{b-a}{n} = \frac{2-0}{4} = .5$. So the intervals are 0 to .5, .5 to 1.0, 1.0 to 1.5, and 1.5 to 2.0. The midpoints of the intervals are .25, .75, 1.25, and 1.75, respectively.

Store the integrand in the calculator as (4 X) ÷ (X + 6). Evaluate at each midpoint, storing the value in a memory location, say M, and adding to it as each successive evaluation is completed. Multiply the sum by .5 (the width of the subintervals).

Store (4 X) ÷ (X + 6)	
.25 → X	Store .25 as X.
Evaluate the expression.	The result is .16.
Ans→ M	Store the result in memory M.
.75→X	Store .75 as X.
Evaluate the expression.	The result is .44.
Ans + M → M	Add the contents of memory M and store the sum back in memory M. The result is .60.

1.25→X	Store 1.25 as X.
Evaluate the expression.	The result is .69.
Ans + M → M	Add the contents of memory M and store the sum back in memory M. The result is 1.29.
1.75→X	Store 1.75 as X.
Evaluate the expression.	The result is .90.
Ans + M → M	Add the contents of memory M to the result and store the sum back in memory M. The result is 2.20.
Ans×.5	Multiply the result by Δx. The result is 1.10.

Exercise Set 12

1. Repeat Example 2 for the derivatives in Problems 21-29 of Textbook Exercise 12-1.
2. Repeat Example 3 for the derivatives in Problems 21-29 of Textbook Exercise 12-1.
3. Graph the derivative and the function with C = 0 on the same set of coordinate axes for Problems 35-56 of Textbook Exercise 12-1.
4. Graph the derivative and the function with C = 0 on the same set of coordinate axes for Problems 73-78 of Textbook Exercise 12-1.
5. Graph the derivative and the function with C = 0 on the same set of coordinate axes for Problems 1-38 of Textbook Exercise 12-2.
6. Graph the derivative and the function with C = 0 on the same set of coordinate axes for Problems 45-62 of Textbook Exercise 12-2.
7. Solve Problems 25-26 of Textbook Exercise 12-3.
8. Solve Problems 27-34 of Textbook Exercise 12-3.
9. Repeat Example 4 finding the age of the bone if 12% was remaining.
10. Repeat Example 4 finding the age of the bone if 8% was remaining.
11. Use a TI-82 or TI-85 to solve Problems 7-22 of Textbook Exercise 12-5.
12. Use a TI-82 or TI-85 to solve Problems 29-46 of Textbook Exercise 12-5.
13. Solve Problems 55-60 of Textbook Exercise 12-5.
14. Solve Problems 65-70 of Textbook Exercise 12-5.

15. Evaluate $\int_{2.58}^{5.36} 4x^2+3e^{.2x} + \frac{3}{x}\, dx$. Use two decimal place accuracy.

16. Evaluate $\int_{2.58}^{5.36} 4x^{-1} + e^{-.2x} - \frac{3}{x^2}\, dx$. Use two decimal place accuracy.

17. Use the midpoint sum to approximate $\int_{-1.2}^{1.2} \frac{2x}{2x + 3}\, dx$ with $n = 4$.

Use two decimal places.

18. Use the midpoint sum to approximate $\int_{3.4}^{6.4} \frac{x}{2x^2 + 3}\, dx$ with $n = 3$.

Use two decimal places.

19. Do Textbook Chapter 12 Group Activity. Use the calculator where appropriate.

Chapter 13

Additional Integration Topics

This chapter contains examples using the calculator to illustrate :

- Finding the area between two curves
- Using the rectangle rule

The graphing calculator features used are:

- Calculation
- Replay key
- Storing a function
- Evaluating a function
- Graphing and viewing window variables
- Tracing and zooming

EXAMPLE 1 ***TEXTBOOK SECTION 13-1***

(A) Find, to two decimal places, the area between the graph of $f(x) = 1.5x^2 - 2.3x + 0.8$ and the x axis over the interval [-1, 1].

(B) Conjecture, then check whether or not the desired area can be found by using a single definite integral.

(C) Conjecture, then check, what the area between the graph of $f(x) = 1.5x^2 - 2.3x + 1.8$ and the x axis over the interval [-1, 1].

Solution (A): Enter the function in the calculator as Y1 as 1.5 X ^ 2 - 2.3 X + .8. Graph using viewing window variables [-10, 10]1 by [-10, 10]1. Using zoom we see that the graph crosses the x axis in two places.

TI-81

Use zoom and trace to find the x intercepts. They are (.53, 0) and (1, 0).

Calculate the definite integrals $\int_{-1}^{.53} 1.5x^2 - 2.3x + 0.8\, dx + \int_{.53}^{1} -(1.5x^2 - 2.3x + 0.8)\, dx$

by storing the antiderivative of the integrand and evaluating. The result is 2.63 + .03 = 2.96.

TI-82 and TI-85

Use the solver capabilities of the calculator to find the x intercepts. (See Appendix Section A-9, B-9, or C-9 of this manual.) Use the calculator to evaluate the two integrals

$$\int_{-1}^{.53} 1.5x^2 - 2.3x + 0.8\, dx + \int_{.53}^{1} -(1.5x^2 - 2.3x + 0.8)\, dx.$$ (See Appendix Section B-15 or C-15 of this manual.) The result is 2.63 + .03 = 2.96.

EXAMPLE 2 ***TEXTBOOK SECTION 13-1***

Find the area bounded by $f(x) = \frac{1}{2}x - 3$ and $g(x) = -x^2 + 1$.

Solution: Store the functions as two separate functions in the calculator as .5 X - 3 and - X ^ 2 + 1 .

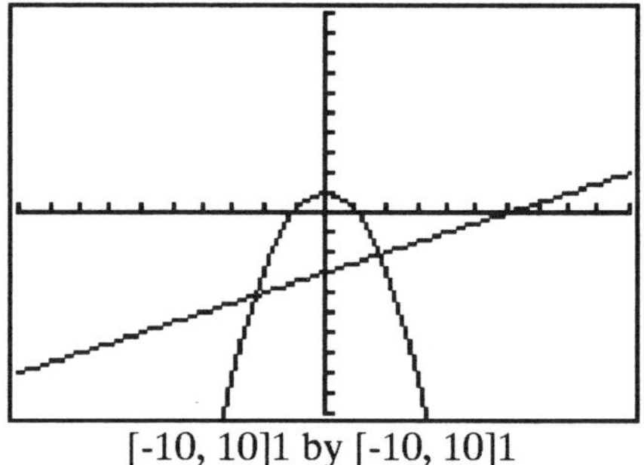

[-10, 10]1 by [-10, 10]1

TI-81

Graph using [-10,10]1 by [-10,10]1. Use zoom and trace to find the points of intersection. They are (1.76, -2.12) and (-2.27, - 4.13).

Evaluate the definite integral by storing the antiderivative as a function in the calculator and evaluating at the limits.

$$\int_{-2.27}^{1.76} (-x^2+1) - (.5x-3)\, dx = -\frac{x^3}{3} + x - .5\left(\frac{x^2}{2}\right) + 3x \Bigg|_{-2.27}^{1.76} = 4.45 - (-6.47) = 10.92$$

TI-82 and TI-85

Graph using [-10, 10]1 by [-10, 10]1. Use equation solving capabilities of the calculator to find the points of intersection. (See Appendix Section B-9 or C-9 of this manual.) The points of intersection are (1.76, -2.12) and (-2.27, -4.13). Use the calculator to evaluate the integral (see Appendix Section B-15, or C-15 of this manual).

EXAMPLE 3 ***TEXTBOOK SECTION 13-2***

Given the normal probability distribution having mean 15.4 and standard deviation 2.3.

(A) Graph using [6, 24]1 by [0, .3].1.

(B) Find $P(8.3 < x < 11.4)$.

(C) Conjecture, then check, how the probability found in Part (B) is effected if the mean was 17 and the standard deviation is 1.5.

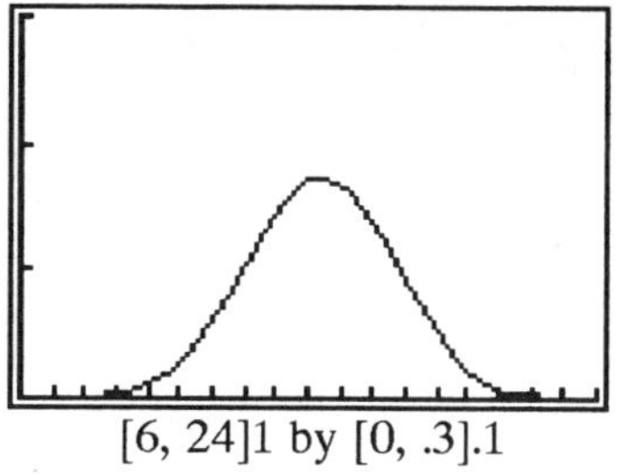
[6, 24]1 by [0, .3].1

Solution (A): The formula for this situation it is:

$$f(x) = \frac{1}{2.3\sqrt{2\pi}} e^{-\frac{(x - 15.4)^2}{2(2.3^2)}}$$

. Store this in the calculator as: (1 ÷ (2.3 × √ (2 × π))) × [e^] (- ((X - 15.4) ^ 2) ÷ (2 × 2.3 ^ 2)). Set the viewing window variables to [6, 24]1 by [0, .3]1. Graph.

Solution (B): We wish to evaluate $\displaystyle\int_{8.3}^{11.4} \frac{1}{2.3\sqrt{2\pi}} e^{-\frac{(x - 15.4)^2}{2(2.3^2)}} dx.$

<u>TI-81</u> The value of this integral can be found by finding the z values that correspond to $x = 8.3$ and $x = 11.4$ and then using the table of normal probability distribution values. The corresponding z values are $z = \frac{8.3\text{-}15.4}{2.3} = -3.09$ and $z = \frac{11.4\text{-}15.4}{2.3} = -1.74$. Hence the value of the integral is .4990-.4591 = .0399.

<u>TI-82</u> Use the built-in numeric integration capability of the calculator. On the home screen enter: [MATH] [9] <fnInt(> [2nd] [Y-Vars] [1] <Function> [1] <Y1> [,] [X,T,θ] [,] [8.3] [,] [11.4] [)] [ENTER]. The result is .03999.

<u>TI-85</u> Use the built-in numeric integration capability of the calculator. On the home screen enter: [2nd] [CALC] [F5] <fnInt(> [2nd] [alpha] [Y] [1] [,] [x-VAR] [,] [8.3] [,] [11.4] [)] [ENTER]. The result is .03999.

Solution (C): The graph will be shifted to the right 1.6 units and will be taller and skinnier since the standard deviation is smaller. Hence the probability will be smaller since the area under the curve is less.

The probability is very close to zero.

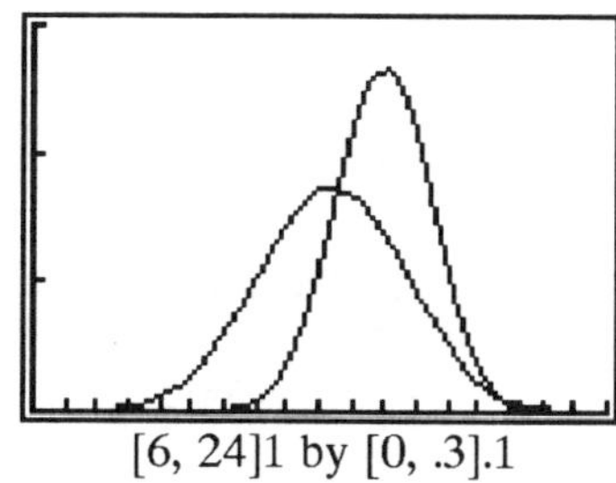
[6, 24]1 by [0, .3].1

Exercise Set 13

1. Solve Problems 7-18 of Textbook Exercise 13-1.
2. Solve Problems 29-44 of Textbook Exercise 13-1.
3. Solve Problems 45-48 of Textbook Exercise 13-1.
4. Solve Problems 49-62 of Textbook Exercise 13-1.
5. Find the area bounded by $f(x) = x - 1$ and $g(x) = -x^3 + 4x^2 - 3$.
6. Find the area bounded by $f(x) = 2x - 1$ and $g(x) = x^3 - 4x^2 + 2$.
7. Solve Problems 1-10 of Textbook Exercise 13-2. Use the numeric integration capability on the TI-82 or TI-85 calculator.
8. Solve Problems 9-12 of Textbook Exercise 13-3. Use the numeric integration capability on the TI-82 or TI-85 calculator.
9. Solve Problems 27-30 of Textbook Exercise 13-3. Use the numeric integration capability on the TI-82 or TI-85 calculator.
10. Solve Problems 41-44 of Textbook Exercise 13-3. Use the numeric integration capability on the TI-82 or TI-85 calculator.
11. Solve Problems 11-16 of Textbook Exercise 13-4. Use the numeric integration capability on the TI-82 or TI-85 calculator.
12. Solve Problems 35-38 of Textbook Exercise 13-4. Use the numeric integration capability on the TI-82 or TI-85 calculator.
13. Graph the integrand and its antiderivative for the integrals of Problems 1-10 of Textbook Exercise 13-4. Use the numeric integration capability on the TI-82 or TI-85 calculator.
14. Graph the integrand and its antiderivative for the integrals of Problems 17-28 of Textbook Exercise 13-4. Use the numeric integration capability on the TI-82 or TI-85 calculator.
15. Do Textbook Chapter 13 Group Activity. Use the calculator where appropriate.

Chapter 14

Multivariable Calculus

This chapter contains examples using the calculator to illustrate:

- Storing and evaluating functions in two or more variables
- Plotting a scatter diagram
- Calculating the least squares line
- Graphing the least squares line
- Predicting using the least squares line

The graphing calculator features used are:

- Calculation
- Storing a function in two or more variables
- Evaluating a function in two or more variables
- Storing bivariate data
- Plotting bivariate data
- Finding the least squares line
- Graphing the least squares line
- Graphing and graph screen limits
- Zoom and trace

EXAMPLE 1 — ***TEXTBOOK SECTION 14-1***

The cost function in dollars of producing surfboards is $C(x, y) = 700 + 70x + 100y$, where x=the number of standard surfboards produced per week and y=the number of competition boards. Evaluate $C(x, y)$ for $x = 10$ and $y = 5$.

Solution: Store the function in the calculator as 700 + 70 X + 100 Y. Store 10 for X and 5 for Y. Recall the function and evaluate. The solution is 1900. It costs $1900 to produce 10 standard surfboards and 5 competition surfboards per week.

EXAMPLE 2 — ***TEXTBOOK SECTION 14-1***

The compound interest formula is $A(P, r, t, n) = P\left(1 + \frac{r}{n}\right)^{nt}$. Evaluate this for P=\$25043, r=8.76%, n=365, and t=5.5.

Solution: Store the function in the calculator as P (1 + R ÷ N) ^ (N × T). Store 25043 for P, .0876 for R, 365 for N, and 5.5 for T. Evaluate the function. The result is $40541.92.

EXAMPLE 3 ***TEXTBOOK SECTION 14-2***

The profit function for the surfboard company in Example 1 above is $P(x, y) = 140x + 200y - 4x^2 + 2xy - 12y^2 - 700$. (A) Find $P_x(15,10)$. (B) Find $P_y(15,10)$

Solution (A): TI-81 The numeric partial derivative cannot be found using the TI-81 calculator. Find the partial derivative algebraically, store it in the calculator, and evaluate it using the calculator. (See Example 1 and 2 above.) The partial derivative with respect to x is $P_x = 140 - 8x + 2y$. Store this as Y1, store 15 as X and 10 as Y. Recall the expression from the Y-VARS list and evaluate. The result is 40. This means that at the production level of 15 standard surfboards and 10 competition surfboards the increase in profit of increasing the production of standard boards by 1 board per week will be approximately $40.

TI-82 Store the function in the calculator as Y1 as [140] [X,T,θ] [+] [200] [Y] [-] [4] [X,T,θ] [^] [2] [+] [2] [X,T,θ] [×] [ALPHA] [Y] [-] [12] [ALPHA] [Y] [^] [2] [-] [700]. Store the value of 10 as Y. Find the first partial with respect to x at x=15 and y=10 by entering [MATH] [8] <nDeriv(> [2nd] [Y-VARS] [1] <Function> [1] <Y1> [,] [X,T,θ] [,] [15] [)]. The result is 40. This means that at the production level of 15 standard surfboards and 10 competition surfboards the increase in profit of increasing the production of standard boards by 1 board per week will be approximately $40.

TI-85 Store the function in the calculator as Y1 as [140] [x-VAR] [+] [200] [2nd] [ALPHA] [Y] [-] [4] [x-VAR] [^] [2] [+] [2] [x-VAR] [×] [2nd] [ALPHA] [Y] [-] [12] [2nd] [ALPHA] [Y] [^] [2] [-] [700]. Store the value of 10 for y. Find the first partial with respect to x at x=15 and y=10 by entering [2nd] [CALC] [F2] <nDer> [2nd] [ALPHA] [Y] [1] [,] [x-VAR] [,] [15] [)]. The result is 40. This means that at the production level of 15 standard surfboards and 10 competition surfboards the increase in profit of increasing the production of standard boards by 1 board per week will be approximately $40.

Solution (B): TI-81 The numeric partial derivative cannot be found using the TI-81 calculator. Find the partial derivative algebraically, store it in the calculator, and evaluate it using the calculator. (See Examples 1 and 2 above.) The partial derivative with respect to y is $P_y = 200 + 2x - 24y$. Store this as Y2, store 15 as X and 10 as Y. Recall the expression from the Y-VARS list and evaluate. The result is -10. This means that at the production level of 15 standard surfboards and 10 competition surfboards, there will be a decrease in profit of approximately $10 if one more competition surfboard is produced per week.

TI-82 Store the function in the calculator as Y1 as 140 X + 200 Y - 4 X ^ 2 + 2 X × ALPHA Y - 12 ALPHA Y ^ 2 - 700 . Store the value of 15 for x. Find the first partial with respect to y at x=15 and y=10 by entering MATH 8 <nDeriv(> 2nd Y-VARS 1 <Function> 1 <Y1> , ALPHA Y , 10) . The result is -10. This means that at the production level of 15 standard surfboards and 10 competition surfboards, there will be a decrease in profit of approximately $10 if one more competition surfboard is produced per week.

TI-85 Store the function in the calculator as Y1 as 140 x-VAR + 200 2nd ALPHA Y - 4 x-VAR ^ 2 + 2 x-VAR × 2nd ALPHA Y - 12 2nd ALPHA Y ^ 2 - 700 . Store the value of 15 for x. Find the first partial with respect to x at x=15 and y=10 by entering 2nd CALC F2 <nDer> 2nd ALPHA Y 1 , 2nd ALPHA Y , 10) . The result is -10. This means that at the production level of 15 standard surfboards and 10 competition surfboards, there will be a decrease in profit of approximately $10 if one more competition surfboard is produced per week.

EXAMPLE 4 ***TEXTBOOK SECTION 14-5***

The following table lists the midterm and final examination scores for 10 students in a calculus course.

Midterm	Final	Midterm	Final
49	65	78	77
53	47	83	81
67	72	85	79
71	76	91	93
74	68	99	99

(A) Find the least squares line for the data given in the table.

(B) Use the least squares line to predict the final examination score for a student who scored 95 on the midterm examination.

(C) Graph the data and the least squares line on the same set of axes.

(D) What score would a student need to get on the midterm examination in order to get an 80 on the final?

Use two decimal places.

Solution (A): Each calculator has its own method for entering data and calculating the least squares line coefficients. (See Appendix Section A-16, B-16, or C-16 of this manual.) Enter the data and find the least squares line. The least squares line for this data is $\hat{y}$ = 13.56 + .83x where the coefficients have been rounded to two decimal places. The calculator will store the least squares line so that you can evaluate it.

Solution (B): Store 95 as X in the calculator. Evaluate the least squares line equation for $\hat{y}$ using the stored equation for the line. The result is 92.27. (See Appendix Section A-16, B-16, or C-16 of this manual on where to find the least squares line equation in the calculator.)

Solution (C): Get the function list and store the least squares line equation by recalling it from the variables list (see Appendix Section A-16, B-16, or C-16 of this manual). We use the viewing window variables [0,100]5 by [-5,100]5. Using -5 as the lower limit on the y axis allows room on the screen for (x, y) values if any tracing is to be done. Otherwise some of the data may be covered up by the numbers displayed at the bottom of the screen.

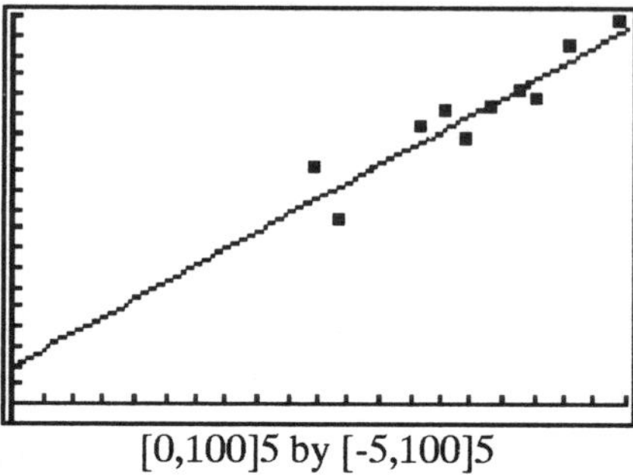

[0,100]5 by [-5,100]5

Solution (D): Use trace and zoom to find the point on the line with y value of 80. (See Appendix Section A-7 & A-9, B-7 & B-9, or C-7 & C-9. The x coordinate of this line is the score a student should get on the midterm examination to predict an 80 on the final examination. The midterm exam score should be 80.37 .

Exercise Set 14

1. Solve Problems 1-8 of Textbook Exercise 14-1.

2. Solve Problems 9-24 of Textbook Exercise 14-1.

3. A manufacturing firm has budgeted $60,000 per month for labor and materials. If $$x$ thousand is spent on labor and $$y$ thousand is spent on materials, and if the monthly output (in units) is given by $N(x, y) = 4x^2+y^2-4xy+2x+3y$ find:

   (A) $N(30, 50)$.
   (B) Graph $N(x, 50)$. Interpret what this means.

4. The research department for a manufacturing company arrived at the following Cobb-Douglas production function for a particular product:
   $N(x, y) = 10x^{0.6}y^{0.40}$ where x is the number of units of labor and y is the number of units of capital required to produce $N(x, y)$ units of the product. Each unit of labor costs $30 and each unit of capital costs $60.

   (A) Find $N(8000, 6000)$.
   (B) Graph $N(x, 6000)$. Interpret what this means.

5. Solve Problems 3-4, 7-8, 11-12, 15-16, 25-28 and 47-48 of Textbook Exercise 14-2.

6. Solve Problems 49-50 by finding the first derivative and solving the equation for y. Graph and find the coordinates of the points of intersection.

7. Solve Problems 1-12 of Textbook Exercise 14-5.

8. Solve Problems 19-20 of Textbook Exercise 14-5.

9. The market research department for a drug store chain chose two summer resort areas to test market a new sun screen lotion packaged in 4 ounce plastic bottles. After a summer of varying the selling price and recording the monthly demand, the research department arrived at the demand table below, where y is the number of bottles purchased per month (in thousands) at x dollars per bottle.

x	5.0	5.5	6.0	6.5	7.0
y	2.0	1.8	1.4	1.2	1.1

(A) Find a demand equation using the method of least squares.
(B) Plot the data and the least squares equation on the same set of axes.
(C) How many bottles are expected to be purchased if the price is $5.25?
(D) What should the price be if we want 1500 bottles to be sold?

10. Repeat Problem 3 using the following data:

x	3.0	3.5	4.0	4.5	6.0
y	2.0	1.8	1.4	1.2	1.1

11. Do Textbook Chapter 14 Group Activity. Use the calculator where appropriate.

NOTES

Appendix A

TI-81 Operations

A-1 Operations

RESET		
1: NO		
2: Reset		
STAT	Bytes	0
PRGM	Bytes	0
Bytes Avail		2400

Press [ON] to turn on the calculator.

OFF

Press [2nd] [ON] to turn off the calculator.

Press [CLEAR] to clear the screen.

Press [2nd] [▲] to make the display darker.

Press [2nd] [▼] to make the display lighter.

Resetting the calculator

Press [2nd] [+] to get the RESET screen . Use the down arrow [▼] to choose 2:Reset and press [ENTER]. The display shows the message **Mem cleared**. Reset clears all functions, programs, modes, and variable values from the calculator's memory. Press [ENTER] to remove message.

Home Screen

The screen where calculations are done is called the Home Screen. You can always get to the

Home

QUIT

Screen by pressing [2nd] [CLEAR]. Note: [QUIT] is above the [CLEAR] key.

In this manual the item above the key will be referenced as if it is on the key. For example, [QUIT], which is above the key, will be referenced from now on as [2nd] [QUIT].

[2nd]

This key must be used to access the calculator functions (written in light blue) above and to the left of a key cap. A new cursor is displayed, [↑].

[ALPHA] and [A-LOCK]

These keys are used to access the calculator functions and memory locations (written in gray) above and to the right of a key cap.

MODE

This key is used to set the operating mode of the calculator. When the highlighted items are on the left the calculator is in default mode. To choose a different mode use the arrow keys to choose the item you want to change to. Then press the ENTER key. The new selection will now be highlighted. Press 2nd QUIT to return to the Home Screen.

Menus

Menus are listed on the screen with a number preceding them. One way to choose an item is to use the arrow keys until the item desired is highlighted. Then press the ENTER key. Another way to choose an item is to just press the number key corresponding to the item you want.

Error Correction

Use the INS key to insert characters.

Use the DEL key to delete characters.

Use the arrow keys ◄ or ► to move to the left or right.

Use the up arrow key ▲ to move up one line or to replay the last executed input.

Use the down arrow key ▼ to move down one line.

A-2 Calculating

Many calculator functions are found above key caps. For example, the square root function is above the key x^2.

The ENTER key must be pressed after the expression has been entered to tell the calculator to calculate.

The subtraction sign and the negative sign are different on the calculator.

The **subtraction sign** is a dark blue key on the lower right side of the calculator. This is used when subtracting one quantity from another.

The **negative sign** (-) is a light gray key in the bottom row of keys on the calculator. This is used when entering a negative quantity.

EXAMPLE 1: Calculate $|-4^2-\sqrt{34}|$

Solution: 2nd ABS ((-) 4 x^2 - 2nd √ 34) ENTER. The result displayed is 21.83095189.

A-3 Plotting Points and Drawing Lines

EXAMPLE 1: (A) Plot the points (-3, 6) and (1,5). (B) Draw the line segment having these points as the endpoints.

Solution (A):

Keystrokes	Explanation
[ZOOM] [6] <Standard>	We first need to set the viewing window dimensions. [ZOOM] [6] <Standard> automatically sets the dimensions to Xmin=-10, Xmax=10, Xscl=1, Ymin=-10, Ymax=10, Yscl=1 and Xres=1. This will be denoted as [-10, 10]1 by [-10, 10]1 in this manual.
[2nd] [DRAW] [1] <ClrDraw> [ENTER]	Get the [DRAW] menu and clear the drawings from the calculator screen.
[2nd] [DRAW] [3] <PT-On(>	Get the [DRAW] menu again. Choose the point on
[(-)] [3] [ALPHA] [,] [6] [)] [ENTER]	option and enter the coordinates of the point with a comma separating them. The point will be plotted.
[2nd] [QUIT]	Return to the home screen.
[2nd] [DRAW] [3] <PT-On(> [ENTER]	Enter the coordinates of the second point.
[1] [ALPHA] [,] [5] [)] [ENTER]	

Solution (B):

Keystrokes	Explanation
[2nd] [QUIT]	Return to the Home Screen.
[2nd] [DRAW] [2] <Line(>	Get the [DRAW] menu and choose the line option. Enter the coordinates of the points separated by commas.
[(-)] [3] [ALPHA] [,] [6] [ALPHA]	The line segment will be drawn between the two points.
[,] [1] [ALPHA] [,] [5] [)] [ENTER]	

A-4 Storing and Evaluating an Expression

EXAMPLE 1: Store the expression $\frac{-b+\sqrt{b^2-4ac}}{2a}$ and evaluate it to solve $3x^2+4x-5=0$.

Solution:

Keystrokes	Explanation
[Y=] [CLEAR]	You will see a list of four function names Y1, Y2, Y3, and Y4. We can store four different functions or algebraic expressions in this list. Then we can recall them and evaluate or graph them.

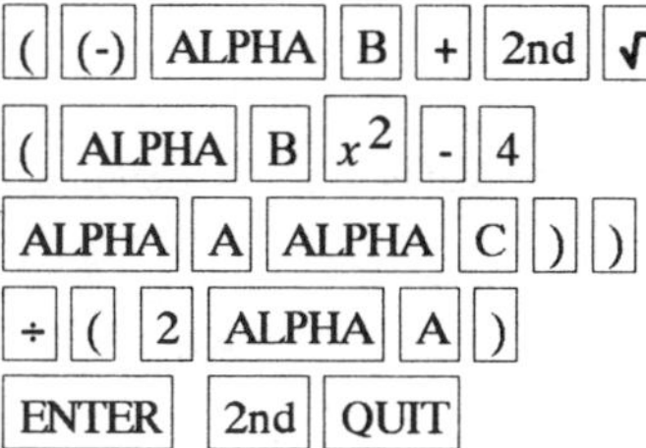

We will use Y1. Enter the expression.

(-B+√(B²-4AC))/(2A)

Note that the B is above and to the right of the key [MATRX]

Get back to the home screen.

[3] [STO▸] [A] [ENTER] [4] [STO▸] [B] [ENTER] [(-)] [5] [STO▸] [C] [ENTER]

Store 3 as A, 4 as B and -5 as C. Using the [STO▸] key automatically puts the calculator in alpha mode.

[2nd] [Y-VARS] [1] <Y1> [ENTER]

Recall the expression and evaluate. The result is .79 to two decimal places. So one solution to this quadratic equation is x=.79.

[Y=] [▶] ... [▶] [-] [ENTER]

Get the function list. Change the + sign to a subtraction sign and evaluate again. The other solution is x=-2.12.

> Throughout this manual, when using menus, the number of the desired selection will be shown followed by what the selection is. Thus, [Y-VARS] [1] <Y1> means to pick the first selection (that is, Y1) off the Y-VARS menu.

EXAMPLE 2: Given $f(x) = \sqrt[3]{x^4+2x}$ and $g(x) = 4^{2x-90}$. Find $f(x) + 3\sqrt{g(x)}$ at $x = 45.2$.

Solution:

[Y=] [CLEAR] or [◀] [ENTER]

Clear the old expression for Y1 by pressing [CLEAR]. Clear or deselect all by placing the cursor over the = sign using the left arrow and pressing enter. Now this function will not graph. Note that the = sign is no longer highlighted.

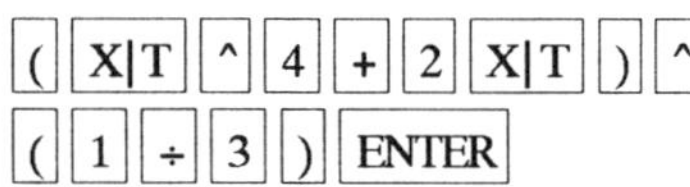

[(] [X|T] [^] [4] [+] [2] [X|T] [)] [^] [(] [1] [÷] [3] [)] [ENTER]

Enter $f(x)$ as Y1.

```
(X^4+2X)^(1/3)
```

Note that the [X|T] key displays the letter X. (The letter T is displayed if parametric mode is being used.)

[4] [^] [(] [2] [X|T] [-] [90] [)] [2nd] [QUIT]

Enter $g(x)$ as Y2.

```
4^(2X-90)
```

[45.2] [STO▸] [X|T] [ENTER]

Store 45.2 as X.

[2nd] [Y-VARS] [1] <Y1> [+] [3] [2nd] [√] [2nd] [Y-VARS] [2] <Y2> [ENTER]

Form $f(x) + 3\sqrt{g(x)}$ and evaluate.

The result is 164.97 to two decimal places.

A-5 Evaluating Inequalities

EXAMPLE 1: Given the inequality $x^2 - 3t < 5x + 9t^2$. Test this inequality for $x = 1.89$ and $t = -4.52$.

Solution:

[1.89] [STO▸] [X|T] [ENTER] [(-)] [4.52] [STO▸] [T] [ENTER]

Store the value of the variables.

X|T ^ 2 - 3 ALPHA T 2nd TEST 5 << > 5 X|T + 9 ALPHA	Enter the expression. `X^2-3T<5X+9T^2`
T ^ 2 ENTER	The result is 1. This means the inequality is true for these values of the variables. If the result was a 0 then the inequality would not have been true.

A-6 Graphing

EXAMPLE 1: Graph $y = 3x^2-6$ and $y=3x^2+1$ on the same graphing screen.

Solution:

Y= CLEAR	Get the function list.
3 X|T ^ 2 - 6 ENTER CLEAR	Enter the functions as Y1 and Y2. You may have to clear the function locations first before entering a new function. To clear just press CLEAR.
3 X|T ^ 2 + 1 ENTER	Clear or deselect any other entered functions. To deselect, place the cursor over the = sign using the left arrow and press enter. Now this function will not graph. Note that the = sign is no longer highlighted.
2nd DRAW 1 <ClrDraw> ENTER	Clear all drawings.
2nd STAT ▶ ▶ 2 <ClrStat> ENTER	Clear the statistical registers.
ZOOM 6 <Standard>	The sixth option on the ZOOM menu will automatically set the graphing screen to have $-10<x<10$ and $-10<y<10$. Also, the functions will be automatically graphed.

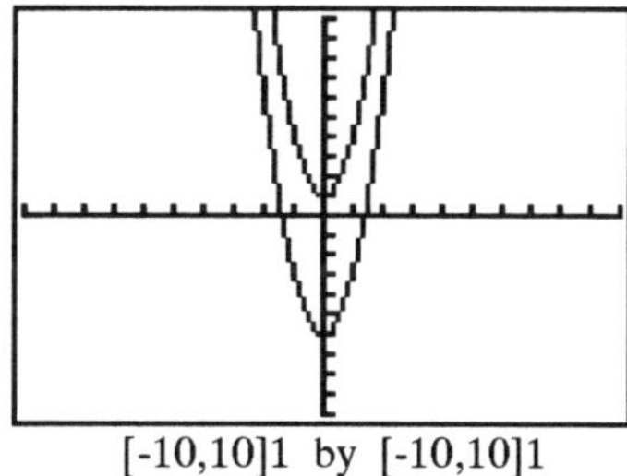

[-10,10]1 by [-10,10]1

> **ZOOM** **6** automatically sets the viewing window variables to $-10<x<10$ and $-10<y<10$ with scale marks every unit for both the x and y axes. The viewing window variables will be denoted in this manual by the notation [-10, 10]1 by [-10, 10]1. The viewing window variables can be set to other values, if desired, by pressing RANGE entering the desired values, and then pressing GRAPH.

> **6** **<Standard>** means to choose the sixth option from the ZOOM menu.

> To deselect a function move the cursor over the = sign and press [ENTER]. You notice the highlighting disappears from the = sign. This means that the function will <u>not</u> graph. You can still evaluate this expression for stored values of the variables. To select the function, place the cursor on the = sign and press [ENTER] again.

EXAMPLE 2: Graph $f(x) = \begin{cases} .5x + 1 & x \le 2 \\ x^2 + 2 & x > 2 \end{cases}$

Solution: The calculator will only graph functions for values of x for which the function is defined. So we write this piecewise-defined function in a form where the pieces are defined only for the values of x that we want.

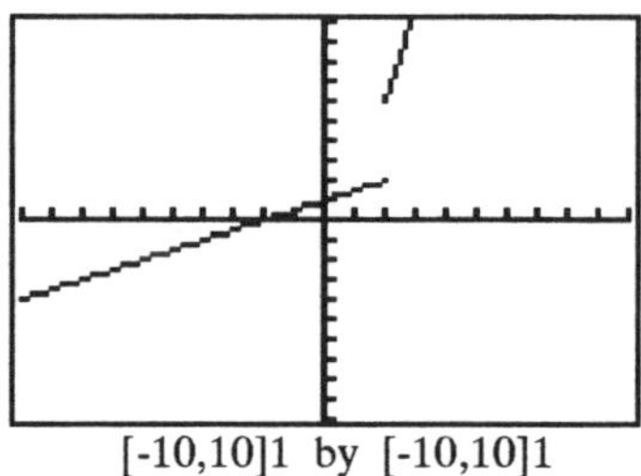
[-10,10]1 by [-10,10]1

Enter the function as two separate functions:

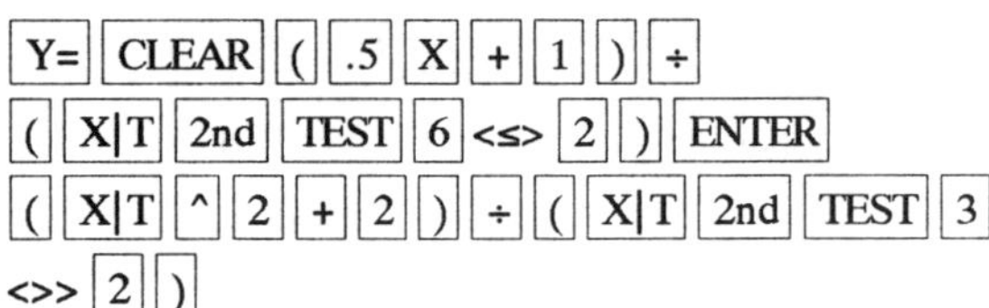

The first function is defined when $x \le 2$ and the second function is defined when $x > 2$.

A-7 TRACE, ZOOM and RANGE

EXAMPLE 1: Find the x intercept for $x>0$ of $y = 3x^2-6$. Use three decimal place accuracy.

Solution: Graph $y = 3x^2-6$ (see Section A-6 Example 1 above).

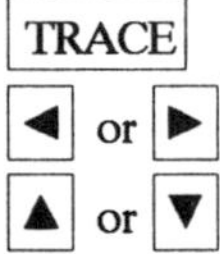

You see a new cursor near the center of the graph screen and the coordinates of that point on the graph displayed at the bottom of the screen.

Use right and left arrow keys and move the cursor along the graph. If we had two or more functions graphed we could use the up and down arrow keys to move between the functions.

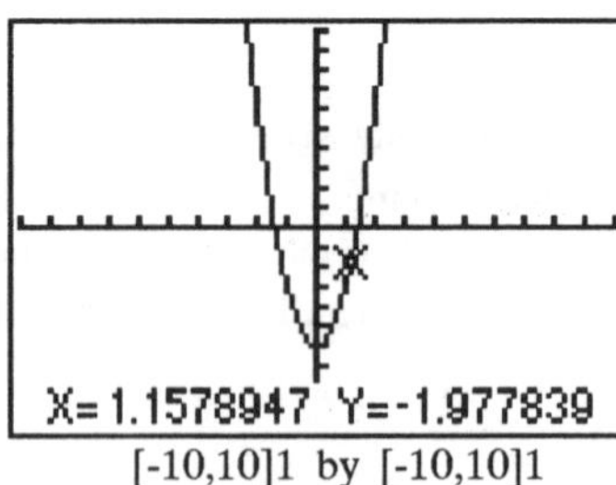

[-10,10]1 by [-10,10]1

We can zoom in on a point to get a closer look. There are three ways to do this:

1. Change the RANGE values.
2. Set the ZOOM FACTORS and ZOOM IN.
3. Use the ZOOM BOX feature of the calculator.

METHOD 1 Set the RANGE

Using trace on the function graphed above we see the x intercept is near 1.5. So we shall set the RANGE closer to that value.

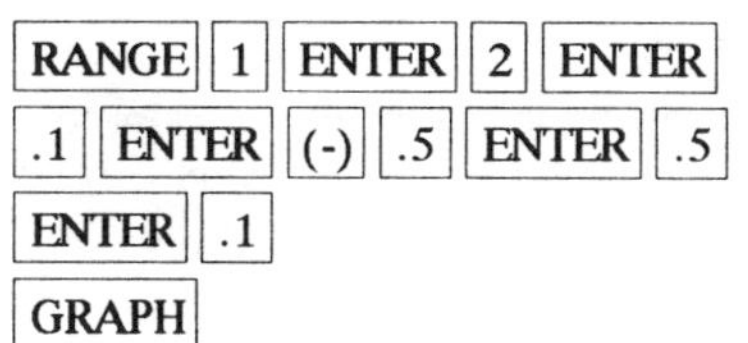

Set the RANGE to [1,2].1 by [-.5,.5].1. Notice we enter the minimum, maximum and scale values for the x axis, and then do the same for the y axis.

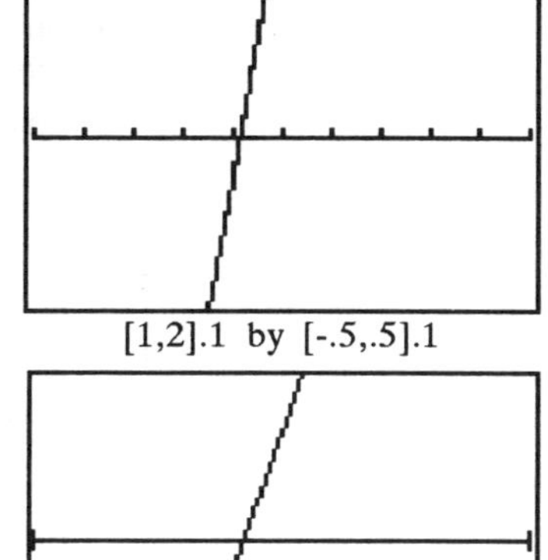

[1,2].1 by [-.5,.5].1

Use trace again to get a better approximation. We see that the x intercept is between 1.41 and 1.42. Set the RANGE to [1.41,1.42].01 by [-.01,.01].001. We now find the x intercept to be 1.414 accurate to three decimal places.

[1.41,1.42].01 by [-.01,.01].001

METHOD 2 Zoom In

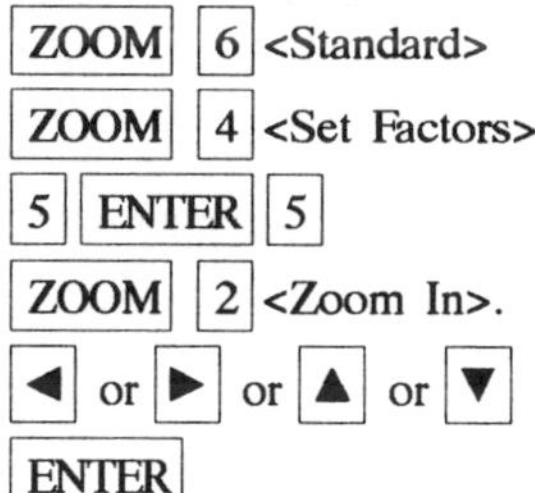

Get the [-10,10]1 by [-10,10]1 display as shown in Section A-6 Example 1.

Set the zoom factors to 5 and 5. After zooming in the new Xmin will be $\frac{1}{5}$ the previous Xmin.

Use the arrow keys repeatedly to get the cursor close to the x intercept, say at x=1.5789474 y=.15873016. The new display will be centered where you placed the cursor.

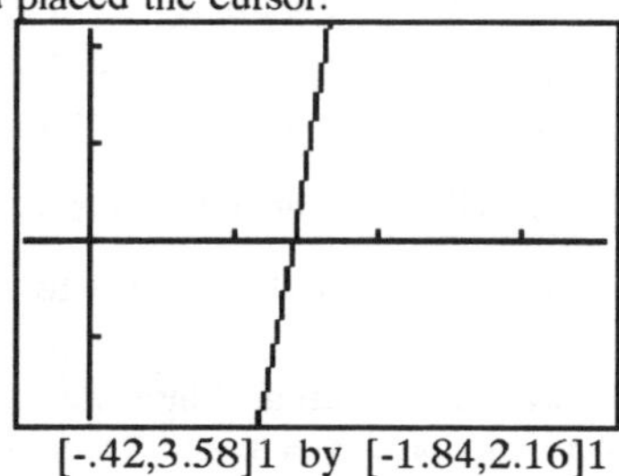

[-.42,3.58]1 by [-1.84,2.16]1

Now use TRACE again to get a new approximation to the x intercept.
Repeat the zooming in process as many times as necessary to get three decimal place accuracy.
The x intercept is 1.414 .

METHOD 3 Zoom Box

Keys	Description
ZOOM 6 <Standard>	Get the [-10,10]1 by [-10,10]1 display as shown in Section A-6 Example 1.
ZOOM 1 <Box>	Get the zoom box option.
◄ or ► or ▲ or ▼ ENTER	Use the arrow keys repeatedly to get the cursor close to but a little above and to the left of the x intercept, say at X=1.1578947 Y=.79365079. Press ENTER to place the upper corner of the zoom box at this point.
◄ or ► or ▲ or ▼ ENTER	Now move the cursor down and to the right of the x intercept, say at X=2 Y=-.4761905. Press ENTER again. The graph will be drawn again with the box as the outside of the new display.

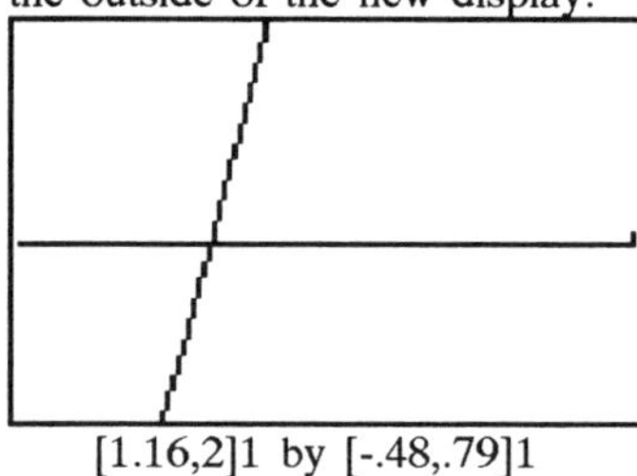

[1.16,2]1 by [-.48,.79]1

Use trace to get an estimate of the x intercept. Use Zoom Box a second time to get an even better approximation of the x intercept. The x intercept is 1.414 .

A-8 Choosing the RANGE Variables

There is no one right way to determine the RANGE that should be used for the graph of a function. Three methods to try are:

1. Graph using the standard screen and then zoom in or zoom out if necessary.
2. Analyze the features of the function and then choose a RANGE.
3. Evaluate the function at several values and then choose a RANGE.

EXAMPLE 1 Graph $y = .03x^3 + .02x^2 - 5x + 10$.

Solution:

METHOD 1 Graph using the standard screen and then zoom out

Enter the function as .03 X ^ 3 + .02 X ^ 2 - 5 X + 10. Press ZOOM 6 to get the [-10,10]1 by [-10,10]1 display. (See Section A-6 of this manual.) We have too small a display to see the behavior of the graph clearly. Press ZOOM 4 and set the zoom factors to XFact=4 and YFact=4. Press ZOOM 3, move the cursor so it is where you want the center of the new screen to be, (we will leave it at the origin for this example) and press ENTER. The result is shown below. Note the double lines as the axes. This is because the Xscl and Yscl values do not change when zoom in is used. Since the scale values for x and y are small compared to the distance between Xmin and Xmax, and Ymin and Ymax, the scale marks appear to run together. You can fix this by choosing a more appropriate scale (say at 5 for Xscl and Yscl for this example).

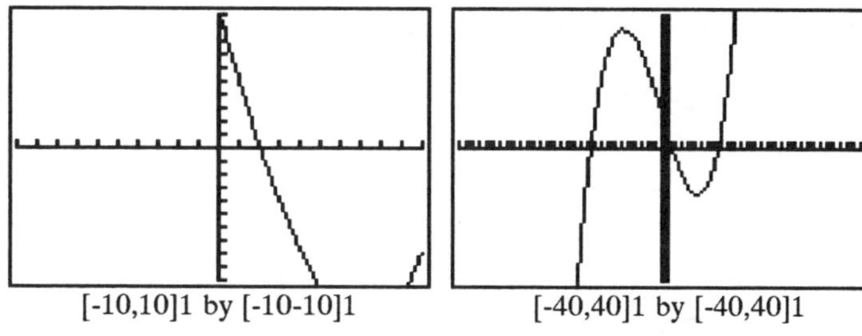

[-10,10]1 by [-10-10]1 [-40,40]1 by [-40,40]1

METHOD 2 Analyze the features of the function and then choose a RANGE

We will analyze the term with highest power on x since this is the dominant term of the function. We see that the leading coefficient is .03. This means that the function will increase 1 unit every time x^3 increases $\frac{1}{.03} \approx 30$ units (x increases $\sqrt[3]{30} \approx 3$ units). A good first choice for the limits on the x axis is Xmax = 10×(unit increase in x) = 30 and Xmin=-30. A good number of scale marks for the axes is 20. Hence the Xscl could be set at $\frac{\text{Xmax-Xmin}}{20} = \frac{30-(-30)}{20} = 3$.

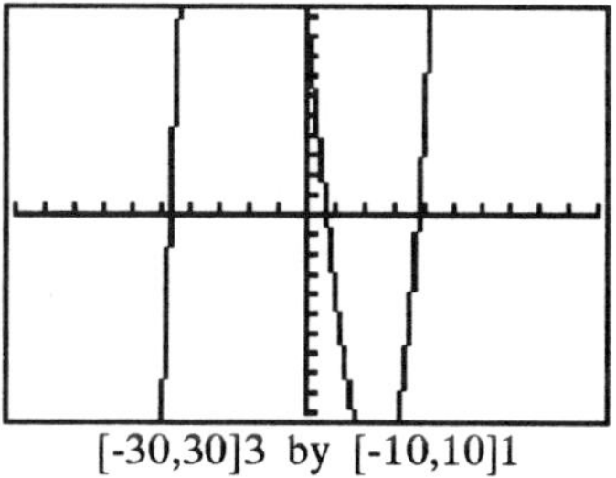

[-30,30]3 by [-10,10]1

The constant term is 10. This means the y intercept is 10. Hence a good first choice for the y axis limits is Ymax=10 and Ymin=-10. The Yscl is then $\frac{\text{Ymax-Ymin}}{20} = \frac{10-(-10)}{20} = 1$. Set the RANGE at [-30,30]3 by [-10,10]1. As can be seen, the RANGE needs to be adjusted further to get a good display.

METHOD 3 Evaluate the function at several values and then choose a RANGE

Evaluate the function by storing a value for x and evaluating the function (see Section A-4). Repeat for each value of x. The choice of x values is arbitrary.

x	$f(x)$
-25	-321.25
-10	32
0	10
10	-8
25	366.25

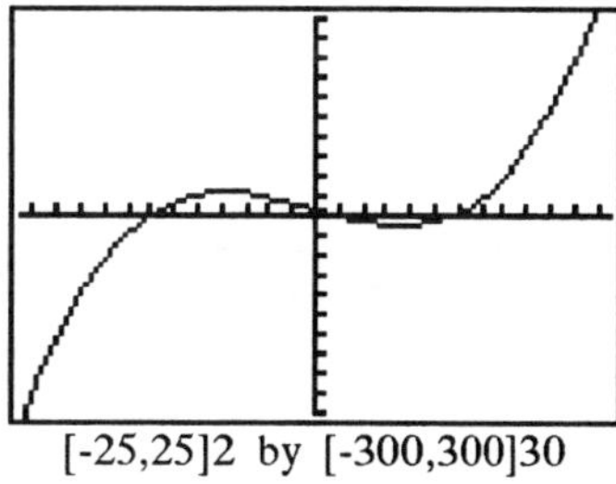

[-25,25]2 by [-300,300]30

Choose the Xscl and Yscl so there are about 20 marks on each axis (see METHOD 2 above). So a good first choice for the RANGE could be [-25,25]2 by [-300,300]30.

A-9 Equations in One Variable

An equation in one variable can be solved graphically in two ways:

1. Algebraically move all terms to one side of the equation. Create a function $f(x)$=(the expression). Graph the function. Use zoom and trace to find the x intercepts. The set of x intercepts is the solution to the equation.

2. Create two functions: $f(x)$ = (the left side of the equation) and $g(x)$ = (the right side of the equation). Graph both functions. Use zoom and trace to find the x value of the points of intersection of the two graphs. The set of x values is the solution to the equation.

EXAMPLE: Solve $\sqrt[3]{x+2} = 4x^2 - 3\sqrt{x+4}$. Use two decimal places.

Solution:

<u>METHOD 1</u> Write the equation as $\sqrt[3]{x+2} - (4x^2 - 3\sqrt{x+4}) = 0$. Create the function $f(x) = \sqrt[3]{x+2} - (4x^2 - 3\sqrt{x+4})$. Enter this function into the calculator and graph.

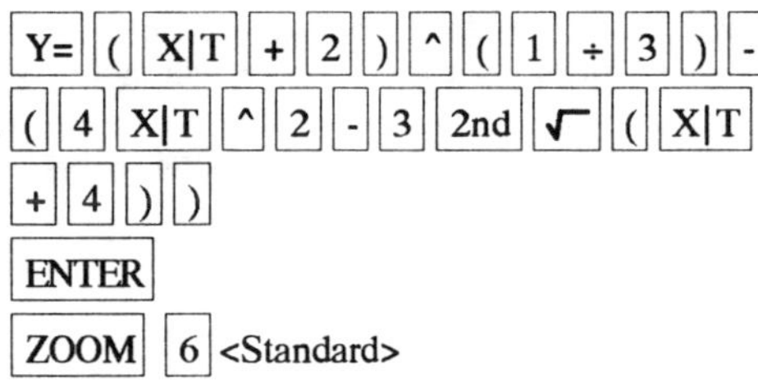

Enter the function and graph.

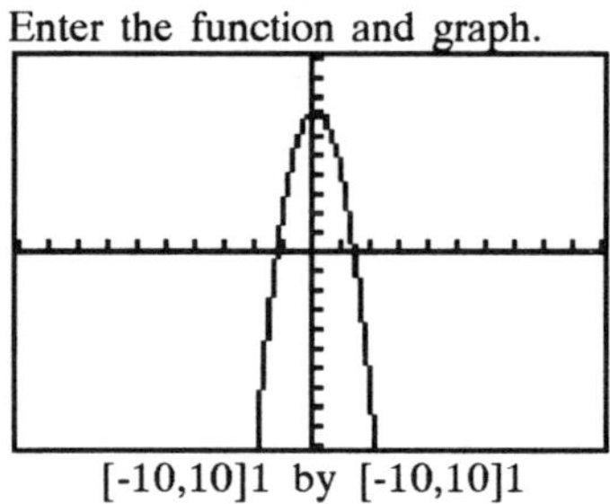

[-10,10]1 by [-10,10]1

We see the function crosses the x axis in two places. Use trace and zoom to find these x values. The solution to this equation is $x = -1.22$ and $x = 1.46$.

<u>METHOD 2</u> Create the two functions $f(x) = \sqrt[3]{x+2}$ and $g(x) = 4x^2 - 3\sqrt{x+4}$. Graph these and find the x values of the intersection points.

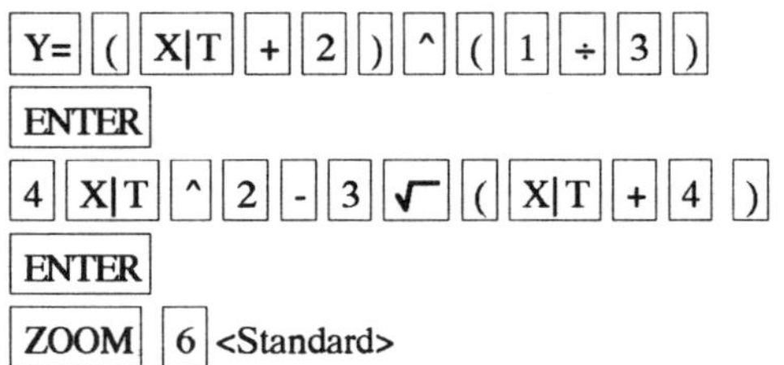

Enter the functions and graph.

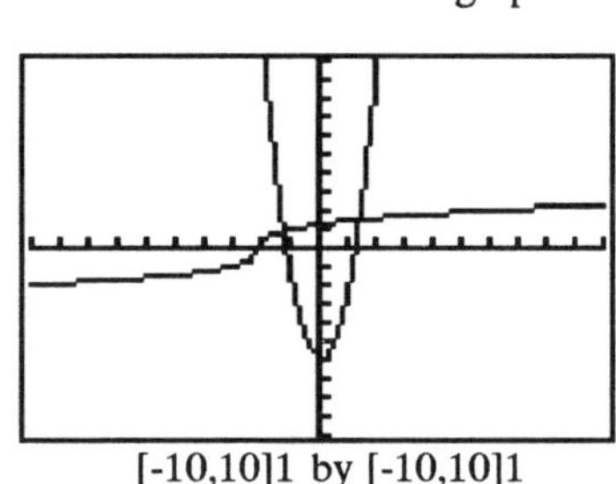

[-10,10]1 by [-10,10]1

We see the functions intersect in two places. Use zoom and trace to find the coordinates of these points. The intersection points are approximately at (-1.22, .92) and (1.46, 1.51). So the solution to this equation is $x = -1.22$ and $x = 1.46$.

<u>A-10 Inequalities in One Variable</u>

An inequality in one variable can be solved graphically in two ways:

1. Algebraically move all terms to one side of the inequality and change the sign to an equal sign. Create a function $f(x)$=(the expression on the left side of the = sign). Graph the function. Use zoom and trace to find the x intercepts. The set of x intercepts are the cutoff values for the solution. Determine the intervals over which the inequality is true and write the solution.

2. Create two functions: $f(x)$ = (the left side of the inequality) and $g(x)$ = (the right side of the inequality). Graph both functions. Use zoom and trace to find the x value of the points of intersection of the two graphs. The set of x coordinates are the cutoff values for the solution. Determine the intervals over which the inequality is true and write the solution.

EXAMPLE: Solve $\sqrt[3]{x+2} < 4x^2 - 3\sqrt{x+4}$. Use two decimal places
Solution:

<u>METHOD 1</u> Write the inequality as: $\sqrt[3]{x+2} - (4x^2 - 3\sqrt{x+4}) < 0$. Change the sign to an equal sign:

$\sqrt[3]{x+2} - (4x^2 - 3\sqrt{x+4}) = 0$. Create the function $f(x) = \sqrt[3]{x+2} - (4x^2 - 3\sqrt{x+4})$, graph, and find the x intercepts. This was done above in Appendix Section A-9 METHOD 1 of this manual. The cutoff values are $x = -1.22$ and $x = 1.46$. Observing the graph we see it is below the x axis for $x < -1.22$ and $x > 1.46$. Thus for these values of x we have $f(x)<0$ and hence

$\sqrt[3]{x+2} < 4x^2 - 3\sqrt{x+4}$. Hence the solution to this inequality can be written as $(-\infty,-1.22)\cup(1.46,\infty)$.

<u>METHOD 2</u> Create the two functions $f(x) = \sqrt[3]{x+2}$ and $g(x) = 4x^2 - 3\sqrt{x+4}$. Graph these and find the x values of the intersection points. This was done above in Appendix Section A-9 METHOD 2 of this manual. The intersection points are at (-1.22, .92) and (1.46, 1.52). Inspecting the graph we see that $f(x) < g(x)$ for $x < -1.22$ and $x > 1.46$. Hence the solution to this inequality can be written as $(-\infty,-1.22)\cup(1.46,\infty)$.

<u>A-11 Permutations, Combinations, Random Numbers</u>

EXAMPLE 1: Find $P_{19,13}$.
Solution:

[19] [MATH] [►] [►] [►] [2] <nPr>
[13] [ENTER]

Enter the numbers and get the built-in function from the MATH menu. The result is displayed in scientific notation as 1.689515283E14.

EXAMPLE 2: Find $C_{19,13}$.
Solution:

[19] [MATH] [►] [►] [►] [3] <nCr>
[13] [ENTER]

Enter the numbers and get the built-in function from the MATH menu. The result is 27132.

EXAMPLE 3: Generate ten random numbers at least 0 and less than 50.
Solution:

[5] [STO►] [ALPHA] [MATH]
[►] [►] [►] [1] <Rand> [ENTER]

We need to store a "seed" number in the storage location RAND. This tells the random number generator where to start.

[MATH] [►] [2] <IPart> [(]
[50] [x] [MATH] [►] [►] [►]
[1] <Rand> [)] [ENTER] [ENTER]
[ENTER] [ENTER] [ENTER] [ENTER]
[ENTER] [ENTER] [ENTER] [ENTER]

RAND generates a number greater than 0 and less than 1. If we multiply by 50 and take the integer part, the result is a number greater than or equal to 0 and less than 50.
The 10 random numbers are:
36 13 6 17 21 ~~21~~ 20 5 31 16 48
Note that repetitions can occur but 21 cannot be used twice.

A-12 Matrices

Given the matrices $A = \begin{bmatrix} 1 & -4 & 6 \\ 2 & -1 & 0 \\ 5 & -3 & 6 \end{bmatrix}$ and $B = \begin{bmatrix} 4 & 15 \\ -2 & 6 \\ 2 & 5 \end{bmatrix}$.

EXAMPLE 1: Find -2.5AB.

Keystrokes	Explanation
MATRX ▶ 1 <[A]>	Choose matrix A and enter the dimensions.
3 ENTER 3 ENTER	Enter the matrix elements.
1 ENTER (-) 4 ENTER	
6 ENTER 2 ENTER	
(-) 1 ENTER 0 ENTER	Choose matrix B and enter the dimensions.
5 ENTER (-) 3 ENTER	Enter the matrix elements.
6 ENTER MATRX	
▶ 2 <[B]>	
3 ENTER 2 ENTER	
4 ENTER 15 ENTER	
(-) 2 ENTER 6 ENTER	Return to the Home Screen. Note that [A] is the matrix A. This is different than A which represents the variable A.
2 ENTER 5	
2nd QUIT	

(-) 2.5 2nd [A] 2nd [B] ENTER

Enter the matrix operations.

The result is $\begin{bmatrix} -60 & -52.5 \\ -25 & -60 \\ -95 & -217.5 \end{bmatrix}$.

EXAMPLE 2: Find A^{-1} using matrix A from Example 1.

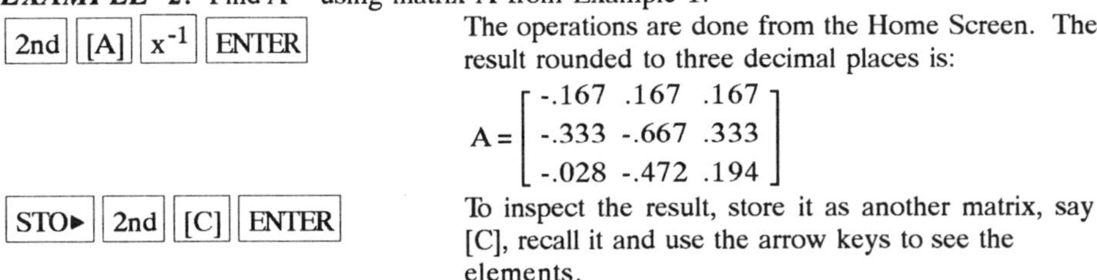

2nd [A] x^{-1} ENTER

The operations are done from the Home Screen. The result rounded to three decimal places is:

$$A = \begin{bmatrix} -.167 & .167 & .167 \\ -.333 & -.667 & .333 \\ -.028 & -.472 & .194 \end{bmatrix}$$

STO▶ 2nd [C] ENTER

To inspect the result, store it as another matrix, say [C], recall it and use the arrow keys to see the elements.

EXAMPLE 3: Solve the system of equations using the inverse of the coefficient matrix. Round answers to two decimal places.

$$\begin{aligned} x - 4y + 6z &= 2.5 \\ 2x - y &= 3.2 \\ 5x - 3y + 6z &= -3.4 \end{aligned}$$

Solution: The method of finding the inverse of a matrix was shown above in Section A-12 Example 2. Enter the constants as a 3x 1 matrix B and find $A^{-1}B$.

Keystrokes	Explanation
MATRX ▶ 2 <[B]>	Enter the matrix dimensions.
3 ENTER 1 ENTER 2.5	Enter the matrix elements.
ENTER 3.2 ENTER (-) 3.4	Return to the Home Screen.
2nd QUIT	Perform the matrix operations.
2nd [A] x^{-1} 2nd [B] ENTER	The result is $\begin{bmatrix} -.45 \\ -4.10 \\ -2.24 \end{bmatrix}$. The solution to the system of equations to two decimal places is x = -.45, y = -4.10, z = -2.24 .

EXAMPLE 4: Find the solution to the system of Example 3 above using Gauss-Jordan reduction. In other words, find the reduced form of the matrix $\begin{bmatrix} 1 & -4 & 6 & 2.5 \\ 2 & -1 & 0 & 3.2 \\ 5 & -3 & 6 & -3.4 \end{bmatrix}$.

Keystrokes	Explanation
MATRX ▶ 1 <[A]> 3	Enter the matrix dimensions.
ENTER 4 ENTER 1 ENTER	Enter the matrix elements.
(-) 4 ENTER 6 ENTER	
2.5 ENTER 2 ENTER	
(-) 1 ENTER 0 ENTER	
3.2 ENTER 5 ENTER	
(-) 3 ENTER 6 ENTER	
(-) 3.4 2nd QUIT	

Some numbers may look like decimals. However upon examining the matrix using the arrow keys you will find they are decimals that are very close to zero. This situation can be eliminated by setting the number of decimal places to a number, say 3 (see Section A-3 of this manual).

Keystrokes	Explanation
MATRX 4 <*Row+(>	We wish to multiply by -2, matrix A, row 1 and add it to row 2. The result will be put in row 2.
(-) 2 ALPHA , 2nd [A]	`*Row+(-2,[A],1,2)`
ALPHA , 1 ALPHA , 2	
) ENTER	The result is $\begin{bmatrix} 1 & -4 & 6 & 2.5 \\ 0 & 7 & -12 & -1.8 \\ 5 & -3 & 6 & -3.4 \end{bmatrix}$.
STO▶ 2nd [A] ENTER	Store the result as matrix A.

Keystrokes	Explanation
[MATRX] [4] <*Row+(> [(-)] [5] [ALPHA] [,] [2nd] [[A]] [ALPHA] [,] [1] [ALPHA] [,] [3] [)] [ENTER]	Now multiply by -5, matrix A, row 1 and add it to row 3. `*Row+(-5,[A],1,3)` The result is $\begin{bmatrix} 1 & -4 & 6 & 2.5 \\ 0 & 7 & -12 & -1.8 \\ 0 & 17 & -24 & -15.9 \end{bmatrix}$.
[STO▸] [2nd] [[A]] [ENTER]	Store the result as matrix A.
[MATRX] [3] <*Row> [1] [÷] [7] [ALPHA] [,] [2nd] [[A]] [ALPHA] [,] [2] [)] [ENTER]	Multiply by 1/7, matrix A, row 2. `*Row(1÷7,[A],2)` The result is $\begin{bmatrix} 1 & -4 & 6 & 2.5 \\ 0 & 1 & -1.71 & -.26 \\ 0 & 17 & -24 & -15.9 \end{bmatrix}$ rounded to two decimal places.
[STO▸] [2nd] [[A]] [ENTER]	Store the result as matrix A.

Continue in this manner until you get $\begin{bmatrix} 1 & 0 & 0 & -.45 \\ 0 & 1 & 0 & -4.10 \\ 0 & 0 & 1 & -2.24 \end{bmatrix}$. The solution to the system of equations is $x = -.45$, $y = -4.10$, $z = -2.24$.

> Note: To swap rows of a matrix use [1] <RowSwap(> from the [MATRX] menu. The command RowSwap([A],2,3) will swap rows 2 and 3 in matrix A.

A-13 Logarithmic, Exponential, and Hyperbolic Functions

EXAMPLE 1: Graph $f(x) = 10^{3.2x}$ for $-.5 < x < .5$.

Solution: Set the RANGE to [-.5,.5].1 by [-10,10]1.

[Y=] [10] [^] [(] [3.2] [X|T] [)]

[GRAPH]

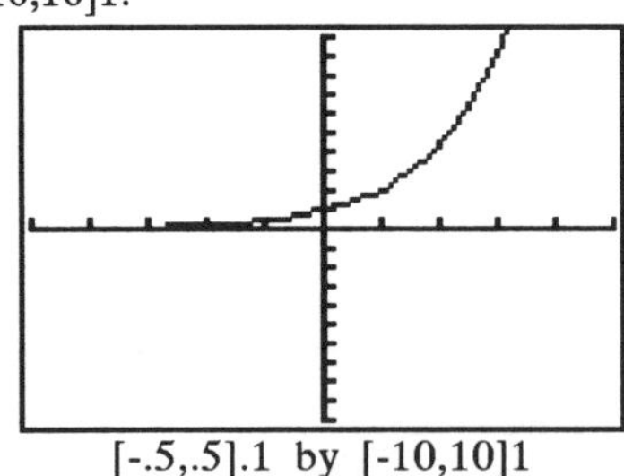

[-.5,.5].1 by [-10,10]1

EXAMPLE 2: Graph $y = \dfrac{e^x + e^{-x}}{2}$.

Solution: This could be entered using [(] [2nd] [e^x] [X|T] [+] [2nd] [e^x] [(-)] [X|T] [)] [÷] [2]. However this is the hyperbolic cosine function. The following uses this built-in function. Recall that [e^x] represents the built-in function above the [LN] key.

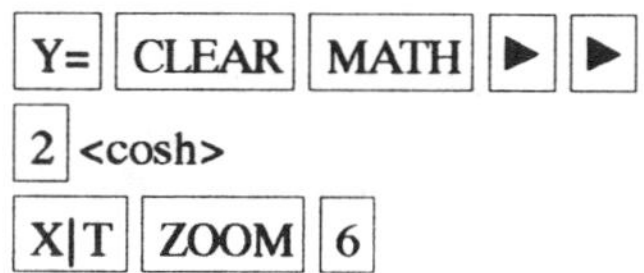

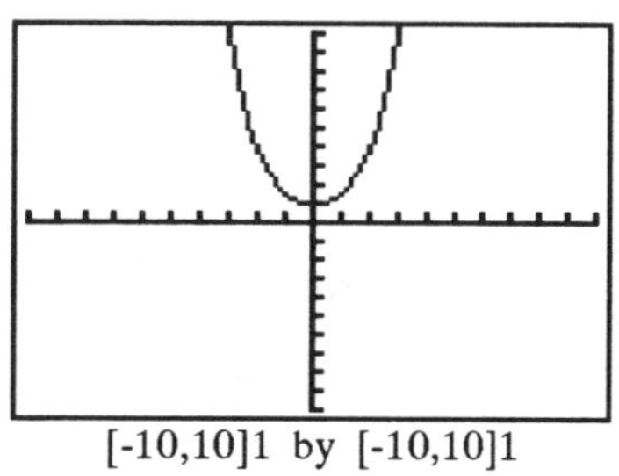

[-10,10]1 by [-10,10]1

EXAMPLE 3: Solve $\ln(x^2+x) = 3x + 2$

Solution:

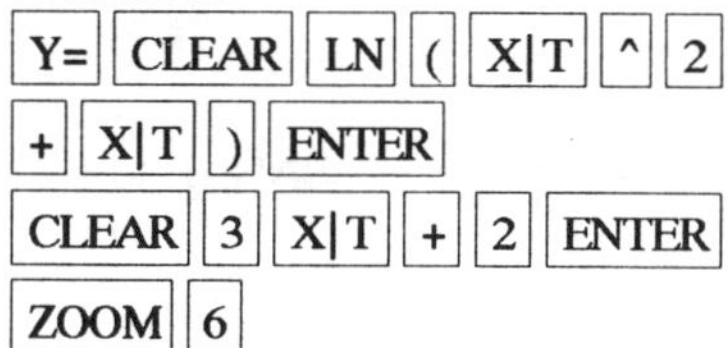

Enter the function $f(x) = \ln(x + x^2)$ as Y1 in the function list.

Enter the function $g(x) = 3x + 2$ as Y2 in the function list.

Graph them on the same graphing screen. Use ZOOM and TRACE to find the x value of the point of intersection. This will be the solution to the equation.

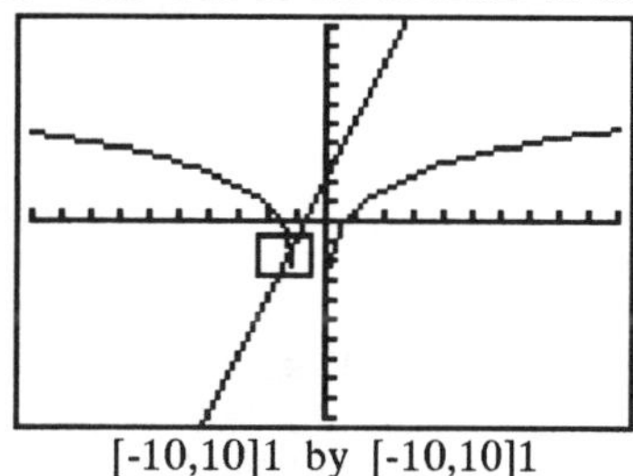

[-10,10]1 by [-10,10]1

The solution is -1.18, rounded to two decimal places.

A-14 Trigonometric Functions

EXAMPLE 1: Evaluate $f(x) = \cos x + \csc x$ at (A) 53°45'18" (B) 78.52 radians

Solution: (A)

Y= CLEAR COS X|T + 1 ÷ SIN X|T ENTER — Store the function.

MODE ▼ ▼ ▶ ENTER — Change the calculator to degree angle measure.

2nd QUIT

53 + 45 ÷ 60 + 18 ÷ 3600 STO▶ X|T — Change the angle to decimal degrees while storing it as the variable X.

2nd Y-VARS 1 <Y1> ENTER — Evaluate the function. The result to two decimal places is 1.83.

Solution:* *(B)

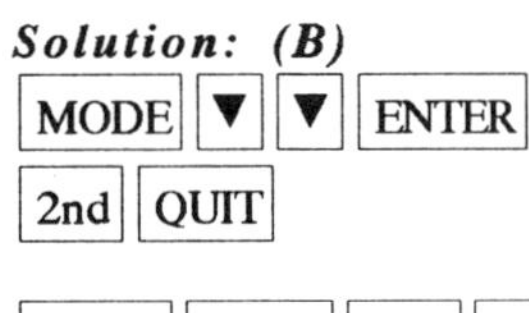

Change the calculator to radian angle measure.
Since the function was stored in part (A) above, we need only to store the angle measure and evaluate the stored function.

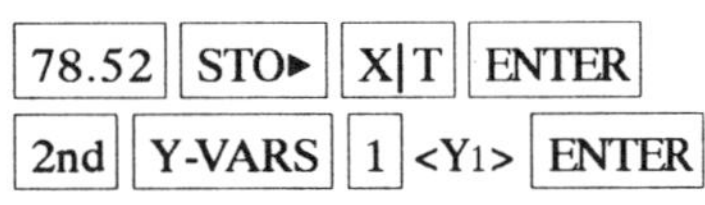

Store 78.52 as X.

Evaluate the function. The result to two decimal places is 49.47.

EXAMPLE 2: Graph $f(x) = \cos x + \csc x$.
Solution: The function was entered in the calculator in Section A-14 Example 1.

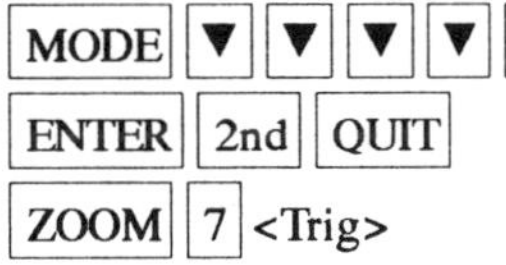

Change the graph mode to Dot. This will eliminate the "vertical lines."

ZOOM 7 <Trig>

This option automatically sets the graph RANGE to [-6.28,6.28]1.57 by [-3,3].25.

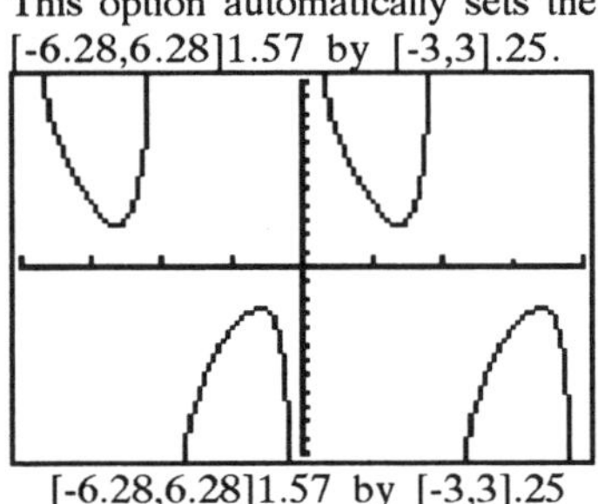

[-6.28,6.28]1.57 by [-3,3].25

A-15 Numeric Differentiation

EXAMPLE 1: Find the equation of the tangent line to $f(x) = \sqrt{x^5 - \sqrt[3]{x}}$ at $x = 3.54$. Use two decimal place accuracy.

Solution: The function and its derivative can be evaluated at this value of x .

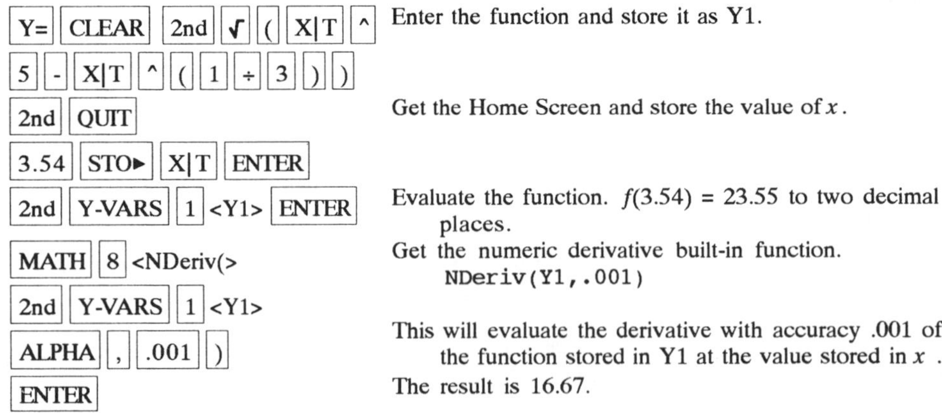

Enter the function and store it as Y1.

Get the Home Screen and store the value of x.

Evaluate the function. $f(3.54) = 23.55$ to two decimal places.

Get the numeric derivative built-in function.

```
NDeriv(Y1,.001)
```

This will evaluate the derivative with accuracy .001 of the function stored in Y1 at the value stored in x .

The result is 16.67.

The equation of the tangent line is $y - 23.55 = 16.67(x - 3.54)$.

A-16 Statistics

EXAMPLE 1: Given the following data find: (A) The mean. (B) The sample standard deviation . (C) A histogram with first class beginning at 0 and width 4. (D) Repeat Parts A and B after deleting the data value 89 from the list. How did eliminating this outlier effect the mean and standard deviation?

23	56	73	12	89	52	45	19	40	57	39
38	18	21	15	37	42	51	38	54	60	

Use one decimal place in answers.

Solution (A) and (B):

[2nd] [STAT] [▶] [▶]
[2] <ClrStat> [ENTER]

Clear the statistical registers.

[2nd] [STAT] [▶] [▶] [1] <Edit>

Get the data entry menu.

[23] [ENTER] [ENTER]
[38] [ENTER] [ENTER]

Enter the data in the x1 position. The y1 position is used for frequency if there is more than one piece of data with the same number.

Etc.

Enter all data. There are 21 values.
If you make a mistake, use the arrow keys to highlight the value and type it in again.

To remove a data value, use the arrow keys to highlight the = sign and press [DEL].

[2nd] [QUIT]

Exit the data entry screen.

[2nd] [STAT] [1] [ENTER]

Get the statistics menu again. Press [1] to get the one variable sample statistics. The mean is 41.86 and the sample standard deviation is 19.82.

Solution (C):

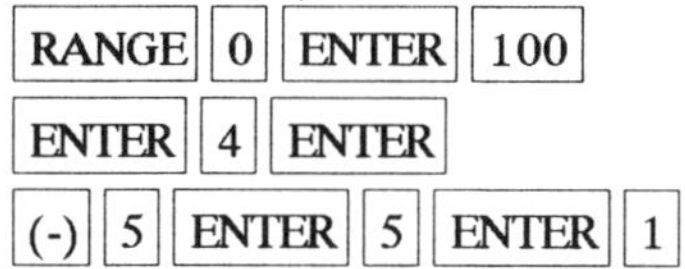

The smallest value is 12 so let xMin=0.
The greatest value is 89. Let us set Xmax=100. The class width is 4, so set Xscl=4. Let Ymin = - 5 to get some room below the histogram. We do not know the frequencies yet so let Ymax=5 for a first try. Let yScl=1.

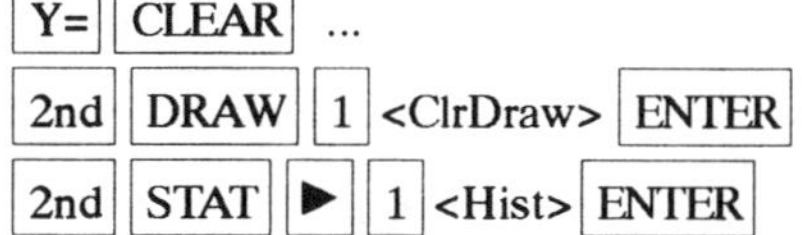

Clear all functions on the function list.

Clear all drawings from the display.

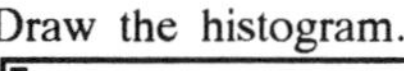

Draw the histogram.

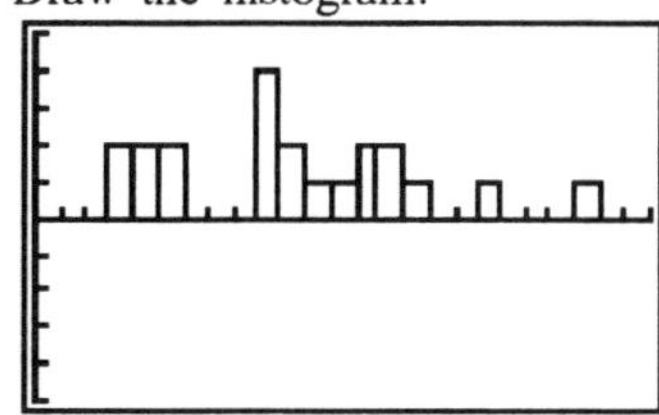

Solution (D):

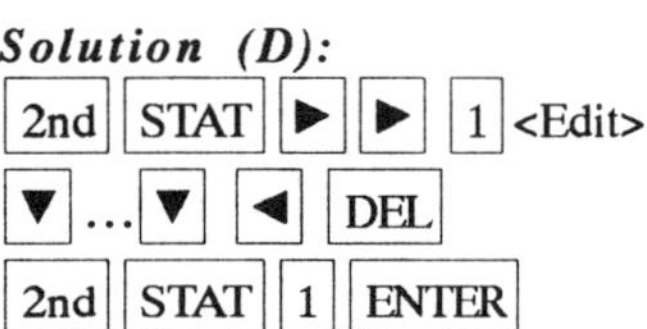

Get the data list and arrow down to the 89. Move the cursor over the equal sign and press DEL.

Get the statistics menu and the one variable sample statistics. The mean is now 39.5 and the sample standard deviation is 17.05. Notice how much influence an outlier has!

EXAMPLE 2: Given the following bivariate data: (A) Find the correlation coefficient. (B) Find the least squares line. (C) Graph the least squares line and the scatter plot on the same graph. (D) Plot the scatter diagram without the least squares line. (E) Sort the data on x and join the points using straight line segments.

x	5	8	10	29	9	24
y	3	4	8	20	3	19

Solution (A):

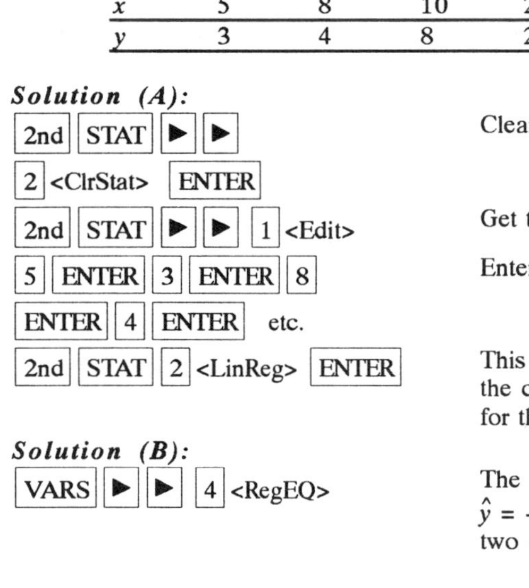

Clear the statistical registers.

Get the data entry menu.

Enter the data.

This gives you the linear correlation coefficient, the constant term and the coefficient on x term for the least squares line. $r = .98$.

Solution (B):

The least squares line equation is displayed. It is $\hat{y} = -1.72 + .79x$ with the coefficients rounded to two decimal places.

Solution (C):

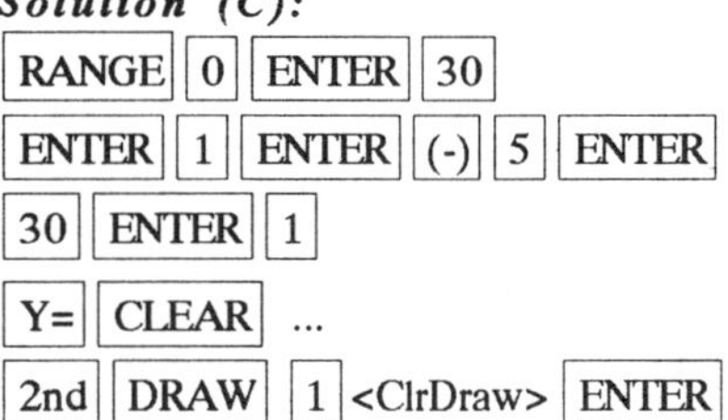

Set the graph RANGE. Observe the data to determine the RANGE values. Set the y value lower than needed so if you trace, the coordinates at the bottom of the screen do not cover up the graph.

Clear all functions on the function list.

Clear all drawings from the display.

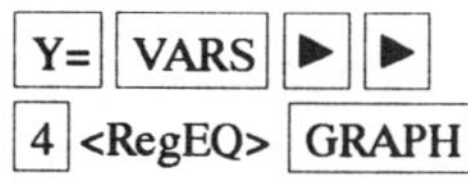

Y= VARS ▶ ▶ 4 <RegEQ> GRAPH

The least squares line equation will be put into the function list so you can graph it. Using GRAPH will graph the line alone.

2nd STAT ▶ 2 <Scatter> ENTER

To get the scatter plot with the line you need to use Scatter from the STAT DRAW menu.

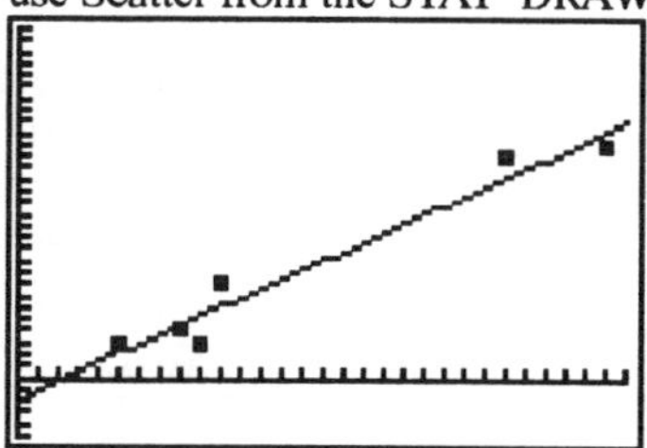

Solution (D):

Y= CLEAR ...

Clear the functions from the list so they won't graph.

2nd STAT ▶ 2 <Scatter> ENTER

Plot the points on the graph.

Solution (E):

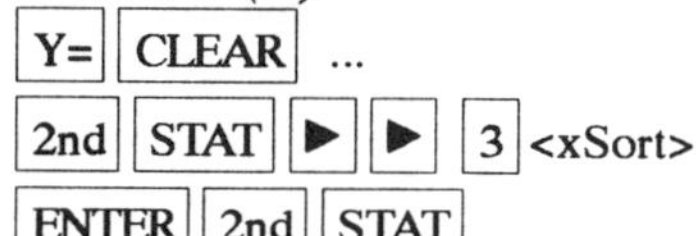

Y= CLEAR ...

Clear the functions from the list so they won't graph.

2nd STAT ▶ ▶ 3 <xSort>

ENTER 2nd STAT

Sort the data on the x variable.

▶ 3 <xyLine> ENTER

This will join the points together with straight line segments in the order they appear in the data list. This is why we sorted the data on x first.

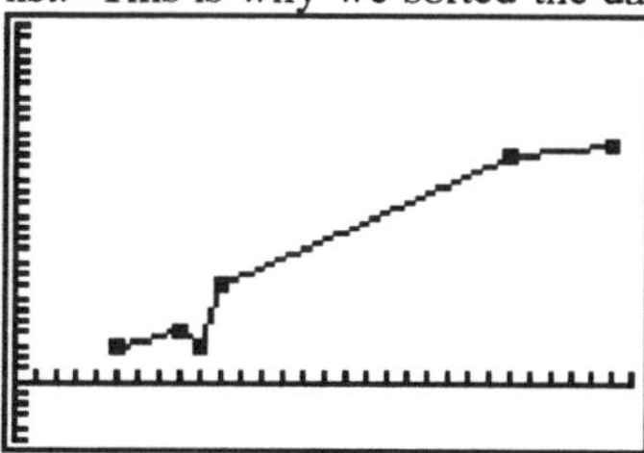

A-17 Setting the Display Mode for Numbers

Numbers can be entered into the calculator in scientific notation, a set number of significant digits, or a fixed number of decimal places.

EXAMPLE 1: Calculate $(9.35417\times 10^4)(-3.64389\times 10^{-6})$ using:

(A) scientific notation (B) two significant digits (C) four decimal place accuracy.

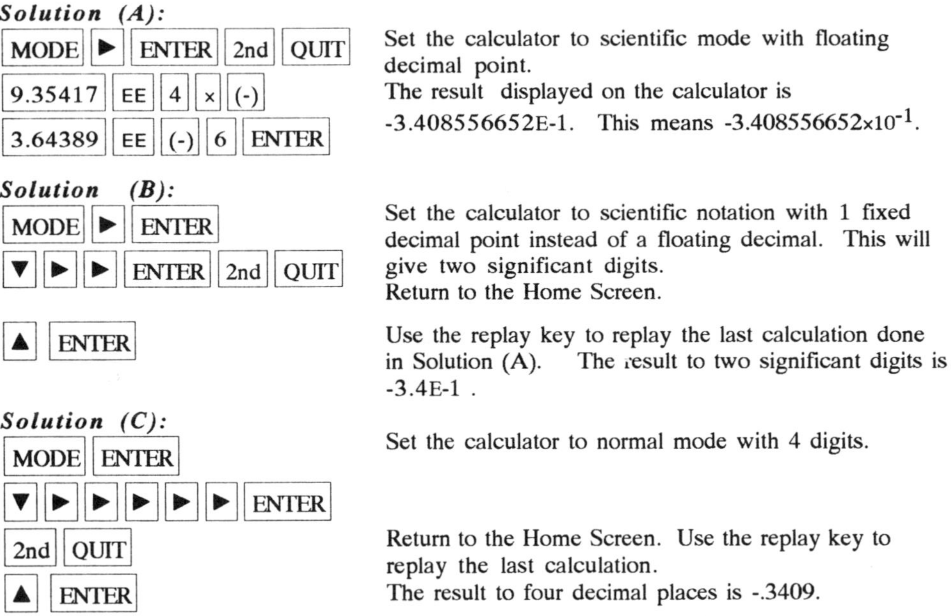

Solution (A):

MODE ▶ ENTER 2nd QUIT — Set the calculator to scientific mode with floating decimal point.

9.35417 EE 4 × (-) 3.64389 EE (-) 6 ENTER — The result displayed on the calculator is -3.408556652E-1. This means $-3.408556652 \times 10^{-1}$.

Solution (B):

MODE ▶ ENTER ▼ ▶ ▶ ENTER 2nd QUIT — Set the calculator to scientific notation with 1 fixed decimal point instead of a floating decimal. This will give two significant digits. Return to the Home Screen.

▲ ENTER — Use the replay key to replay the last calculation done in Solution (A). The result to two significant digits is -3.4E-1 .

Solution (C):

MODE ENTER ▼ ▶ ▶ ▶ ▶ ▶ ENTER — Set the calculator to normal mode with 4 digits.

2nd QUIT ▲ ENTER — Return to the Home Screen. Use the replay key to replay the last calculation. The result to four decimal places is -.3409.

You may now wish to return your calculator to its default settings (all items on the mode screen having the left item highlighted). The examples throughout this manual are worked using the default settings unless otherwise noted.

Appendix B

TI-82 Operations

B-1 Operations

MEMORY
1: Check RAM..
2: Delete...
3: Reset...

Press [ON] to turn on the calculator.

Press [2nd] [+] to get the MEMORY screen (shown at the right).

Choose 3:Reset... by pressing [3] or use the down-arrow key [▼] to highlight 3:Reset... and press [ENTER].

RESET MEMORY
1: No
2: Reset

The display shows the RESET MEMORY .

Choose 2: Reset or use the down-arrow key to highlight 2:Reset... and press [ENTER]. The screen may look blank. This is because the contrast setting was also reset.

Mem cleared should appear on the screen.

If it doesn't, press [2nd] and then hold the [▲] (to increase the contrast) or [▼] (to decrease the contrast) until you see it on the screen.

Press [CLEAR] to clear the screen.

Press [2nd] [OFF] to turn the calculator off.

[2nd]

Press this key to access the operation above and to the left of a key. The operations above the key are in a light blue color on the face of the calculator. Note that a flashing up arrow [↑] appears on the screen after [2nd] key is pressed.

> In this manual, the functions on the face of the calculator above a key will be referred to in square boxes just as if the function was printed on the key cap. For example, [ANS] is the function above the [(-)] key.

[ALPHA] and [A-LOCK]

The operation above and to the right of a key is accessed by first pressing [ALPHA] or [A-LOCK]. A flashing [A] is displayed as the cursor on the screen after the [ALPHA] key is pressed. [A-LOCK] locks the calculator into alpha mode until the [ALPHA] is pressed again.

Home Screen

Calculations are done and commands are entered on the Home Screen. You can always get to this screen (aborting any calculations in progress) by pressing

QUIT
[2nd] [MODE]. From here on, this will be referred to as [2nd] [QUIT] in this appendix.

[MODE]

When you press [MODE] you see the screen shown below. The highlighted items are active.

To change an item select it using the arrow keys and press [ENTER] to activate the selection.

Type of notation for display of numbers.	Normal Sci Eng
Number of decimal places displayed.	Float 0123456789
Type of angle measure.	Radian Degree
Function, parametric, polar or sequence graphing.	Func Par Pol Seq
Connected/not connected plotted points on graphs.	Connected Dot
Graphs functions separately or all at once.	Sequential Simul
Allows a full screen or split screen to be used.	FullScreen Split

Menus

The TI-82 Graphics calculator uses menus for selection of specific functions. The items on the menus are identified by numbers followed by a colon. There are two ways to choose menu items:

1. Press the number corresponding to the menu item.
2. Use the arrow keys to highlight the selection and then pressing [ENTER].

In this manual the menu items will be referred to using the key to be pressed followed by the meaning of the menu. For example, on the ZOOM menu, [1] <ZBox> refers to the first menu item ZBox.

Editing

It is easy to correct errors when entering data into the calculator by using the arrow keys, [DEL], and [INS] keys.

[◄] or [►]	Moves the cursor to the left or right one position.
[▲]	Moves the cursor up one line.
[▼]	Moves the cursor down one line.
[DEL]	Deletes one or more characters at the cursor position.
[2nd] [INS]	Inserts one or more characters at the cursor position.
[2nd] [ENTRY]	Replays the last executed line of input.

B-2 Calculating

EXAMPLE 1: Calculate $|-4^2 - \sqrt{34}|$

Enter numbers and functions in the calculator in the same order as you would read an expression. In the examples that follow, do not press [ENTER] unless specifically instructed to do so. Enter all the keystrokes without pressing the ENTER key until [ENTER] is displayed in the example.

The subtraction sign and the negative sign are different.

> The **subtraction sign** is a dark blue key on the lower right side of the calculator. This is used when subtracting one quantity from another.
>
> The **negative sign** (-) is a light gray key in the bottom row of keys on the calculator. This is used when entering a negative quantity.

Solution: [2nd] [QUIT] [2nd] [ABS] [(] [(-)] [4] [x^2] [-] [2nd] [√] [34] [)] [ENTER]. The result displayed is 21.83095189.

B-3 Plotting Points and Drawing Lines

EXAMPLE 1: (A) Plot the points (-3, 6) and (1,5). (B) Draw the line segment having these points as the endpoints.

Solution (A):

Keystrokes	Explanation
[Y=] [CLEAR] [▼] [CLEAR] etc.	Clear all functions from the function list.
[2nd] [QUIT]	Set the viewing window dimensions.
[ZOOM] [6] <ZStandard> [CLEAR]	[ZOOM] [6] <ZStandard> automatically sets the dimensions to Xmin=-10, Xmax=10, Xscl=1, Ymin=-10, Ymax=10, and Yscl=1. This will be denoted as [-10, 10]1 by [-10, 10]1 in this manual. Press [CLEAR] to get back to the home screen.
[2nd] [DRAW] [1] <ClrDraw> [ENTER]	Get the [DRAW] menu and clear the drawings from the calculator screen.
[2nd] [DRAW] [▶] [1] <PT-On(> [(-)] [3] [,] [6] [)] [ENTER]	Get the [DRAW] menu again. Choose the point on option and enter the coordinates of the point with a comma separating them. The point will be plotted.
[2nd] [QUIT] [2nd] [DRAW] [▶] [1] <PT-On(> [1] [,] [5] [)] [ENTER]	Enter the coordinates of the second point and plot.

Solution (B):

Keystrokes	Explanation
[2nd] [QUIT]	Return to the Home Screen.
[2nd] [DRAW] [2] <Line(>	Get the [DRAW] menu and choose the line option.
[(-)] [3] [,] [6] [,]	Enter the coordinates of the points separated by commas.
[1] [,] [5] [)] [ENTER]	The line segment will be drawn between the two points.

B-4 Storing and Evaluating an Expression

EXAMPLE 1: Store the expression $\frac{-b+\sqrt{b^2-4ac}}{2a}$ and evaluate it to solve $3x^2+4x-5=0$.

Solution:

Keystrokes	Explanation
[Y=] [CLEAR]	You will see a list of function names Y1, Y2, Y3, etc.
[(] [(-)] [ALPHA] [B] [+] [2nd] [√]	We can store ten different functions or algebraic expressions in this list.

(ALPHA B x^2 - 4 ALPHA A × ALPHA C)) ÷ (2 ALPHA A)	Then we can recall them and evaluate or graph them. We will use Y1. Press these keys to enter the expression. `(-B+√(B²-4AC))/(2A)`
ENTER 2nd QUIT	Get back to the home screen.
3 STO▸ ALPHA A ENTER 4 STO▸ ALPHA B ENTER (-) 5 STO▸ ALPHA C ENTER	Store 3 as A, 4 as B and -5 as C.
2nd Y-VARS 1 <Function> 1 <Y1> ENTER	Recall the expression and evaluate. The result is .79 to two decimal places. So one solution to this quadratic equation is x=.79.
	Get the function list.
Y= ▶ ... ▶ - ENTER etc.	Now change the addition sign to a subtraction sign and evaluate again. The other solution is x=-2.12.

> Throughout this manual, when using menus, the number of the desired selection will be shown followed by what the selection is. Thus, Y-VARS 1 <Function> means to pick the first selection (that is, Function) off the Y-VARS menu.

EXAMPLE 2: Given $f(x) = \sqrt[3]{x^4+2x}$ and $g(x) = 4^{2x-90}$. Find $f(x) + 3\sqrt{g(x)}$ at $x = 45.2$.

Solution:

Y= CLEAR	Clear the old expression for Y1 by pressing CLEAR
(X,T,θ ^ 4 + 2 X,T,θ) ^ (1 ÷ 3) ENTER	Enter $f(x)$ as Y1. `(X^4+2X)^(1/3)` Note that the X,T,θ key displays the letter X. (The letter T is displayed if parametric mode is being used, and θ is displayed if in polar graphing mode.)
Y= CLEAR	Clear the function stored as Y2.
4 ^ (2 X,T,θ - 90) 2nd QUIT	Enter $g(x)$ as Y2. `4^(2X-90)`
45.2 STO▸ X,T,θ ENTER	Store 45.2 as X.
2nd Y-VARS 1 <Function> 1 <Y1> + 3 2nd √ 2nd	Form $f(x) + 3\sqrt{g(x)}$ and evaluate.
Y-VARS 1 <Function> 2 <Y2> ENTER	The result is 164.97 to two decimal places.

B-5 Evaluating Inequalities

EXAMPLE 1: Given the inequality $x^2 - 3t < 5x + 9t^2$. Test this inequality for $x = 1.89$ and $t = -4.52$.

Solution:

1.89 STO▸ X,T,θ ENTER — Store the value of the variables.

(-) 4.52 STO▸ ALPHA

T ENTER — Enter the expression.

X^2-3T<5X+9T^2

X,T,θ ^ 2 - 3 ALPHA

T 2nd TEST 5 << > 5

X,T,θ + 9 ALPHA

T ^ 2 ENTER — The result is 1. This means the inequality is true for these values of the variables. If the result was a 0 then the inequality would not have been true.

B-6 Graphing

EXAMPLE 1: Graph $y = 3x^2-6$ and $y=3x^2+1$ on the same graphing screen.

Solution:

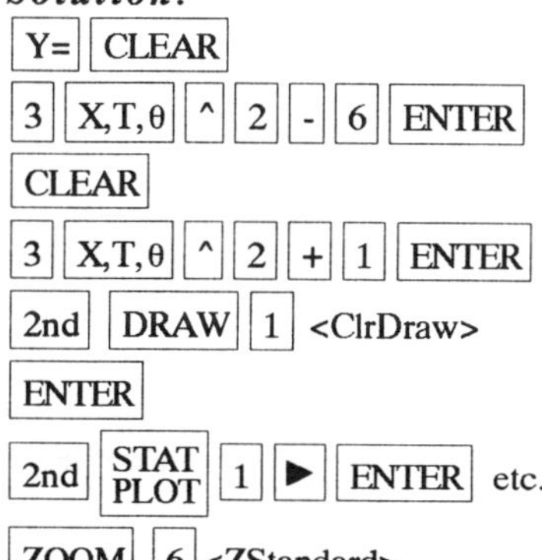

Get the function list and clear the existing function for Y1. Enter the first function as Y1. Now clear the second function location and enter the second function as Y2.

Clear or deselect all others by placing the cursor over the = sign using the left arrow and pressing enter. Now this function will not graph. Note that the = sign is no longer highlighted.

Clear drawings.

Turn off all statistical plots.

ZOOM 6 <ZStandard>

The 6th option on the ZOOM menu will automatically set the graphing screen to have $-10 < x < 10$ and $-10 < y < 10$. Also, the functions will be automatically graphed.

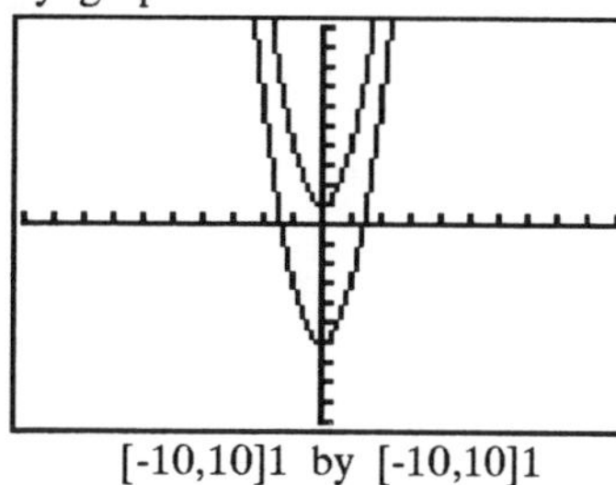

[-10,10]1 by [-10,10]1

| ZOOM | 6 | automatically sets the graphing screen limits to $-10 < x < 10$ and $-10 < y < 10$ with scale marks every unit for both the x and y axes. The graph screen limits will be denoted in this manual by the notation [-10,10]1 by [-10,10]1. The graph screen limits can be set to other values, if desired, by pressing | WINDOW | and entering the desired values, then pressing | GRAPH |.

| 6 | **<ZStandard>** means to choose the sixth option from the ZOOM menu.

The function will not graph if you move the cursor over the = sign and press | ENTER |. You notice the highlighting disappears from the = sign. You can still evaluate this expression for stored values of the variables. To select the function, place the cursor on the = sign and press | ENTER | again.

EXAMPLE 2: Graph $f(x) = \begin{cases} .5x + 1 & x \le 2 \\ x^2 + 2 & x > 2 \end{cases}$

Solution: The calculator will only graph functions for values of x for which the function is defined. So we write this piecewise-defined function in a form where the pieces are defined only for the values of x that we want. Enter the function as two separate functions with (.5 X + 1) ÷ (X ≤ 2) as Y1 and (X ^ 2 + 2) ÷ (X > 2) as Y2. The first function is defined when $x \le 2$ and the second function is defined when $x > 2$. (See Appendix Section B-5 on how to get the inequality symbols from the | TEST | menu, and see Appendix Section B-6 on how to store functions.)

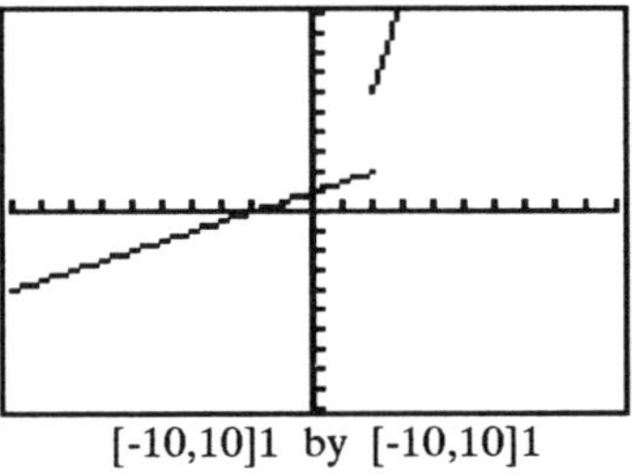

[-10,10]1 by [-10,10]1

B-7 TRACE, ZOOM and WINDOW Variables

First graph $y = 3x^2-6$ on the standard graphing screen (see Appendix Section B-6 Example 1 above).

| TRACE |

| ◄ | or | ► |

| ▲ | or | ▼ |

When you push | TRACE | you see a new cursor near the center of the graph screen and the coordinates of that point on the graph displayed at the bottom of the screen.

Use right and left arrow keys and move the cursor along the graph. If we had two or more functions graphed, we could use the up and down arrow keys to move between the functions.

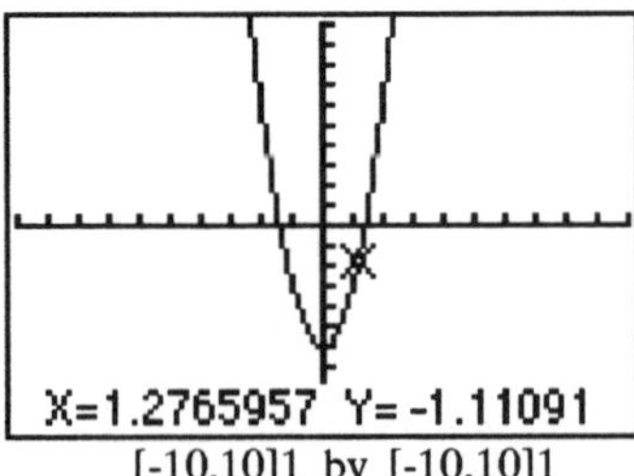

[-10,10]1 by [-10,10]1

We can zoom in on a point to get a closer look. There are three ways to do this:

1. Change the WINDOW dimensions.
2. Set the ZOOM FACTORS and ZOOM IN.
3. Use the ZOOM BOX feature of the calculator.

EXAMPLE 1: Find the x intercept for $x>0$ of $y = 3x^2-6$. Use three decimal place accuracy.

<u>METHOD 1 Set the WINDOW dimensions.</u>
Using trace on the function graphed above we see the x intercept is near 1.5. So we shall set the graph window close to that value.

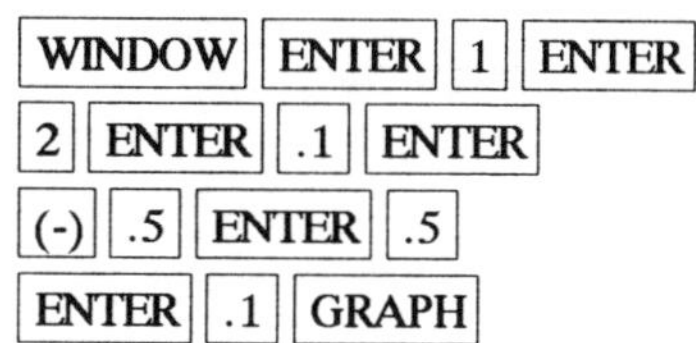

Set the WINDOW to [1,2].1 by [-.5,.5].1. Notice we enter the minimum, maximum and scale values for the x axis, and then do the same for the y axis.

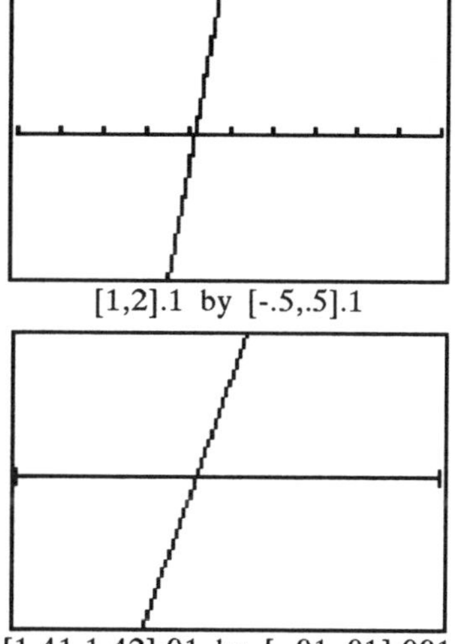

[1,2].1 by [-.5,.5].1

Use trace again to get a better approximation. We see that the x intercept is between 1.41 and 1.42. Set the WINDOW to [1.41,1.42].01 by [-.01,.01].001. We now find the x intercept to be 1.414 accurate to three decimal places.

[1.41,1.42].01 by [-.01,.01].001

<u>METHOD 2 Zoom In</u>

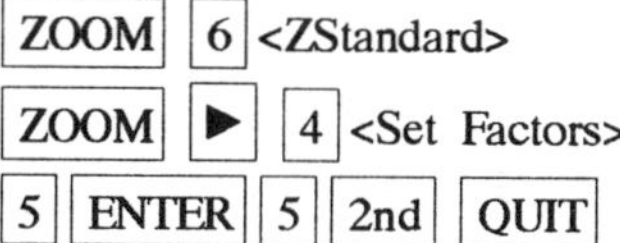

Get the [-10,10]1 by [-10,10]1 display as shown in Section B-6 Example 1. Choose the MEMORY from the ZOOM menu. Choose 4 <Set Factors>.

Set the zoom factors to 5 and 5. After zooming in the new Xmin will be $\frac{1}{5}$ the previous Xmin.

ZOOM 2 <Zoom In>.
◄ or ► or ▲ or ▼
ENTER

Use the arrow keys repeatedly to get the cursor close to the x intercept, say at x=1.4893617 y=.65459482. The new display will be centered where you placed the cursor.

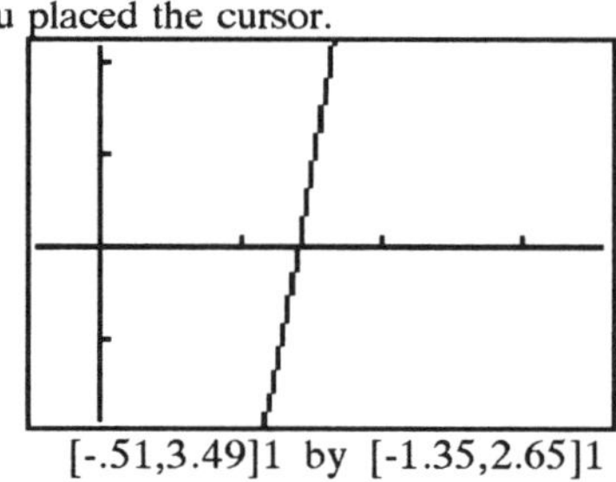

[-.51,3.49]1 by [-1.35,2.65]1

Now use TRACE again to get a new approximation to the x intercept.
Repeat the zooming in process as many times as necessary to get three decimal place accuracy. The x intercept is 1.414 .

<u>METHOD 3 Zoom Box</u>

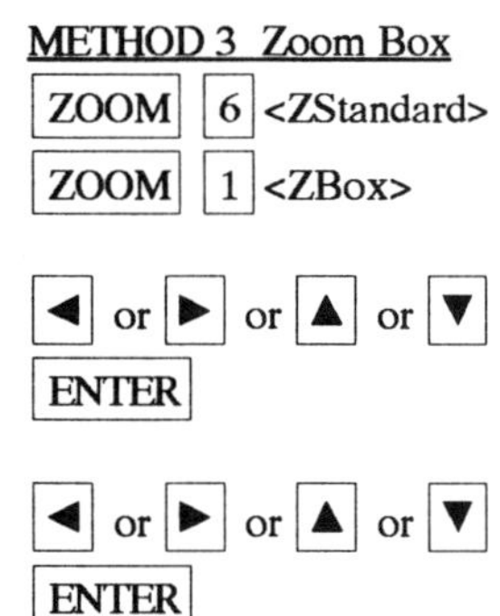

ZOOM 6 <ZStandard>
ZOOM 1 <ZBox>

Get the [-10,10]1 by [-10,10]1 display as shown in Section B-6 Example 1.
Get the zoom box option.

◄ or ► or ▲ or ▼
ENTER

Use the arrow keys repeatedly to get the cursor close to but a little above and to the left of the x intercept, say at X=1.0638298 Y=.64516129. Press ENTER to place the upper corner of the zoom box at this point.

◄ or ► or ▲ or ▼
ENTER

Now move the cursor down and to the right of the x intercept, say at X= 1.7021277 Y=-.3225806. Press ENTER again. The graph will be drawn again with the box as the outside of the new display.

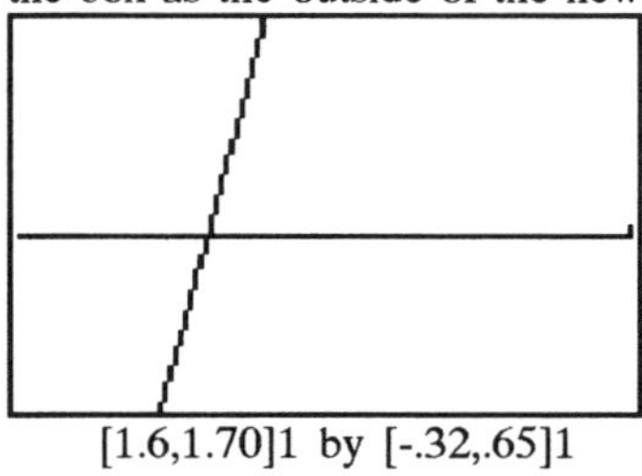

[1.6,1.70]1 by [-.32,.65]1

Use trace to get an estimate of the x intercept. Use Zoom Box a second time to get an even better approximation of the x intercept. The x intercept is 1.414 .

B-8 Choosing the WINDOW Variables

There is no one right way to determine the window variables that should be used for the graph of a function. Three methods to try are:

1. Graph using the standard screen and then zoom in or zoom out if necessary.
2. Analyze the features of the function and then choose the viewing window dimensions.
3. Evaluate the function at several values and then choose the viewing window dimensions.

EXAMPLE 1 Graph $y = .03x^3 + .02x^2 - 5x + 10$.

Solution:

METHOD 1 Graph using the standard screen and then zoom out

Enter the function as .03 X ^ 3 + .02 X ^ 2 - 5 X + 10. Press [ZOOM] [6] to get the [-10,10]1 by [-10,10]1 display. (See Section B-6 of this manual.) We have too small a display to see the behavior of the graph clearly. Press [ZOOM] [▶] [4] and set the zoom factors to XFact=4 and YFact=4. Press [ZOOM] [3], move the cursor so it is where you want the center of the new screen to be, (we will leave it at the origin for this example) and press [ENTER]. The result is shown below. Note the double lines as the axes. This is because the Xscl and Yscl values do not change when zoom in is used. Since the scale values for x and y are small compared to the distance between Xmin and Xmax, and Ymin and Ymax, the scale marks appear to run together. You can fix this by choosing a more appropriate scale (say at 5 for Xscl and Yscl for this example).

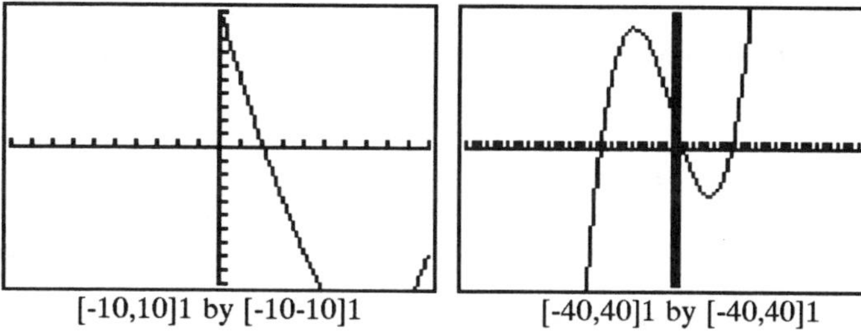

[-10,10]1 by [-10-10]1 [-40,40]1 by [-40,40]1

METHOD 2 Analyze the features of the function and then choose a RANGE

We will analyze the term with highest power on x since this is the dominant term of the function. We see that the leading coefficient is .03. This means that the function will increase 1 unit every time x^3 increases $\frac{1}{.03} \approx 30$ units (x increases $\sqrt[3]{30} \approx 3$ units). A good first choice for the limits on the x axis is Xmax = 10×(unit increase in x) = 30 and Xmin=-30. A good number of scale marks for the axes is 20. Hence the Xscl could be set at $\frac{\text{Xmax-Xmin}}{20} = \frac{30-(-30)}{20} = 3$.

The constant term is 10. This means the y intercept is 10. Hence a good first choice for the y axis limits is Ymax=10 and Ymin=-10. The Yscl is then $\frac{\text{Ymax-Ymin}}{20} = \frac{10-(-10)}{20} = 1$. Set the viewing window variables at [-30,30]3 by [-10,10]1. As can be seen, the variables needs to be adjusted further to get a good display.

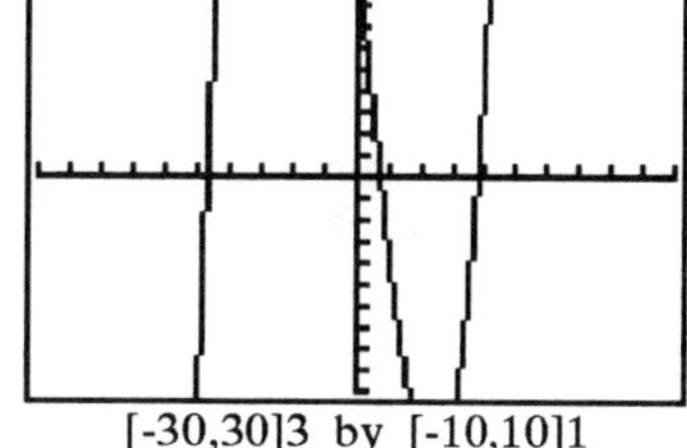

[-30,30]3 by [-10,10]1

METHOD 3 Evaluate the function at several values and then choose the viewing window dimensions

Evaluate the function by storing a value for x and evaluating the function (see Section B-4). Repeat for each value of x. The choice of x values is arbitrary.

x	$f(x)$
-25	-321.25
-10	32
0	10
10	-8
25	366.25

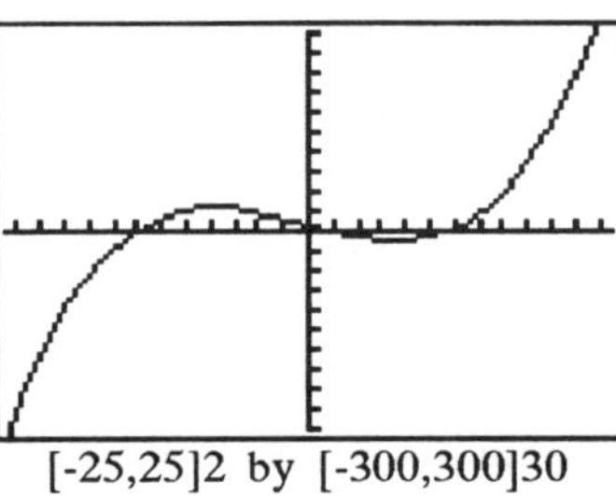
[-25,25]2 by [-300,300]30

Choose the Xscl and Yscl so there are about 20 marks on each axis (see METHOD 2 above). So a good first choice for the WINDOW could be [-25,25]2 by [-300,300]30.

B-9 Equations in One Variable

An equation in one variable can be solved graphically in two ways:

1. Algebraically move all terms to one side of the equation. Create a function $f(x)$=(the expression). Graph the function. Use zoom and trace to find the x intercepts. The set of x intercepts is the solution to the equation.
2. Create two functions: $f(x)$ = (the left side of the equation) and $g(x)$ = (the right side of the equation). Graph both functions. Use zoom and trace to find the x value of the points of intersection of the two graphs. The set of x values is the solution to the equation.

EXAMPLE: Solve $\sqrt[3]{x+2} = 4x^2 - 3\sqrt{x+4}$. Use two decimal places.

Solution:

METHOD 1 Write the equation as $\sqrt[3]{x+2} - (4x^2 - 3\sqrt{x+4}) = 0$. Create the function $f(x) = \sqrt[3]{x+2} - (4x^2 - 3\sqrt{x+4})$. Enter this function into the calculator and graph.

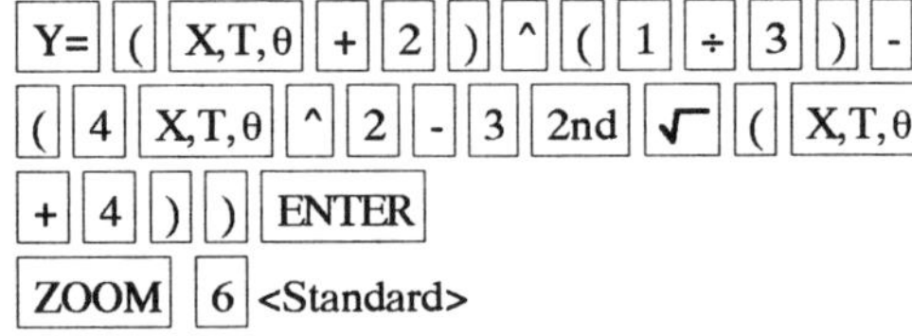

Enter the function and graph.

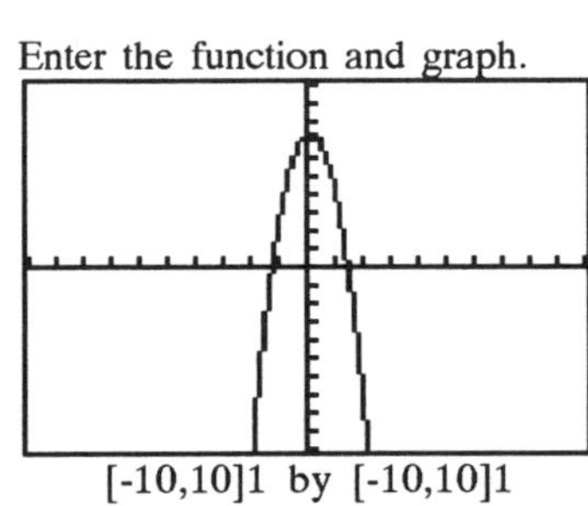
[-10,10]1 by [-10,10]1

We see the function crosses the x axis in two places. Use trace and zoom to find these x values. The solution to this equation is x = -1.22 and x = 1.46.

METHOD 2 Create the two functions $f(x) = \sqrt[3]{x+2}$ and $g(x) = 4x^2 - 3\sqrt{x+4}$. Graph these and find the x values of the intersection points.

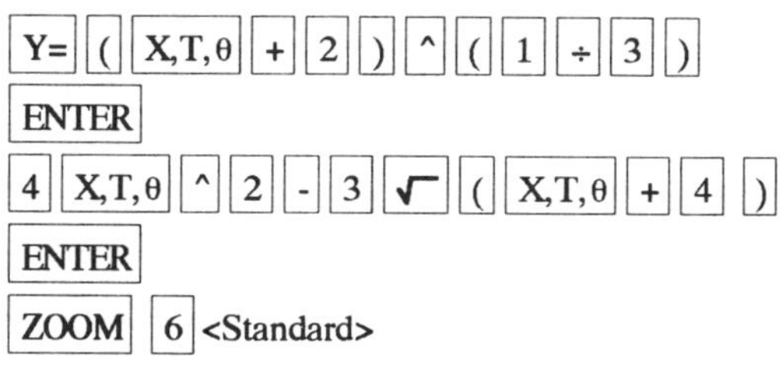

Enter the functions and graph.

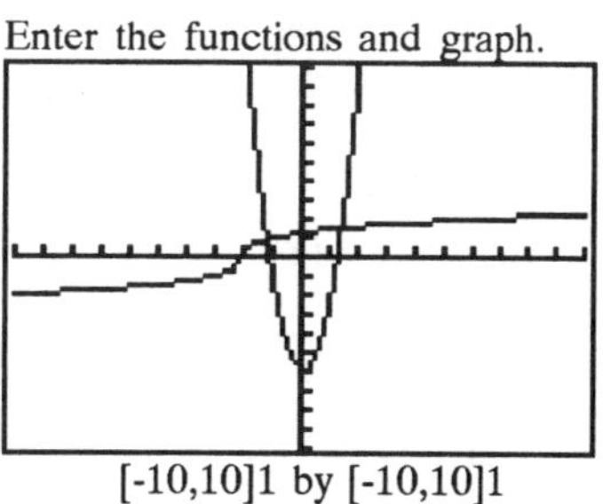

[-10,10]1 by [-10,10]1

We see the functions intersect in two places. Use zoom and trace to find the coordinates of these points. The intersection points are approximately at (-1.22, .92) and (1.46, 1.51). So the solution to this equation is $x = -1.22$ and $x = 1.46$.

B-10 Inequalities in One Variable

An inequality in one variable can be solved graphically in two ways:

1. Algebraically move all terms to one side of the inequality and change the sign to an equal sign. Create a function $f(x)$=(the expression). Graph the function. Use zoom and trace to find the x intercepts. The set of x intercepts are the cutoff values for the solution. Now determine the intervals over which the inequality is true and write the solution. Pick a value either greater or less than the x intercept. Test the inequality using the method shown in Appendix Section B-5.

2. Create two functions: $f(x)$ = (the left side of the inequality) and $g(x)$ = (the right side of the inequality). Graph both functions. Use zoom and trace to find the x value of the points of intersection of the two graphs. The set of x coordinates are the cutoff values for the solution. Now determine the intervals over which the inequality is true and write the solution. Pick a value either greater or less than the x intercept. Test the inequality using the method shown in Appendix Section B-5.

EXAMPLE: Solve $\sqrt[3]{x+2} < 4x^2 - 3\sqrt{x+4}$. Use two decimal places
Solution:

METHOD 1 Write the inequality as: $\sqrt[3]{x+2} - (4x^2 - 3\sqrt{x+4}) < 0$. Change the sign to an equal sign: $\sqrt[3]{x+2} - (4x^2 - 3\sqrt{x+4}) = 0$. Create the function $f(x) = \sqrt[3]{x+2} - (4x^2 - 3\sqrt{x+4})$, graph, and find the x intercepts. This was done above in Appendix Section B-9 METHOD 1 of this manual. The cutoff values are $x = -1.22$ and $x = 1.46$. Observing the graph we see it is below the x axis for $x < -1.22$ and $x > 1.46$. Thus for these values of x we have $f(x)<0$ and hence $\sqrt[3]{x+2} < 4x^2 - 3\sqrt{x+4}$. Hence the solution to this inequality can be written as $(-\infty,-1.22)\cup(1.46,\infty)$.

METHOD 2 Create the two functions $f(x) = \sqrt[3]{x+2}$ and $g(x) = 4x^2 - 3\sqrt{x+4}$. Graph these and find the x values of the intersection points. This was done above in Appendix Section B-9 METHOD 2 of this manual. The intersection points are at (-1.22, .92) and (1.46, 1.52). Inspecting the graph we see that $f(x) < g(x)$ for $x < -1.22$ and $x > 1.46$. Hence the solution to this inequality can be written as $(-\infty,-1.22)\cup(1.46,\infty)$.

B-11 Permutations and Combinations

EXAMPLE 1: Find $P_{19,13}$.

Solution:

[19] [MATH] [►] [►] [►] [2] <nPr>

[13] [ENTER]

Enter the numbers and get the built-in permutation function from the MATH menu. The result is displayed in scientific notation as 1.689515283E14.

EXAMPLE 2: Find $C_{19,13}$.

Solution:

[19] [MATH] [►] [►] [►] [3] <nCr>

[13] [ENTER]

Enter the numbers and get the built-in function from the MATH menu. The result is 27132.

EXAMPLE 3: Generate ten random numbers at least 0 and less than 50.

Solution:

[5] [STO►] [MATH] [►] [►] [►]

[1] <rand> [ENTER]

We need to store a "seed" number in the storage location RAND. This tells the random number generator where to start.

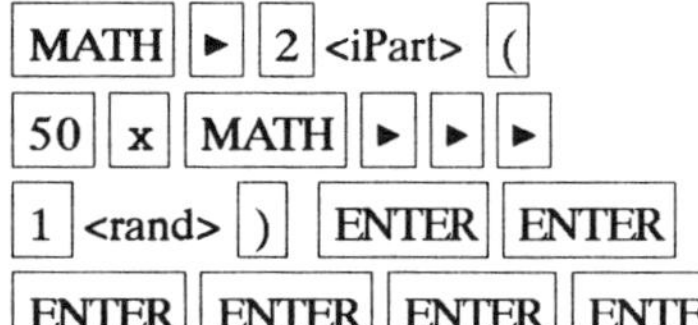

[MATH] [►] [2] <iPart> [(]

[50] [x] [MATH] [►] [►] [►]

[1] <rand> [)] [ENTER] [ENTER]

RAND generates a number greater than 0 and less than 1. If we multiply by 50 and take the integer part, the result is a number greater than or equal to 0 and less than 50.

[ENTER] [ENTER] [ENTER] [ENTER]

[ENTER] [ENTER] [ENTER] [ENTER]

The 10 random numbers are:

36 13 6 17 21 ~~21~~ 20 5 31 16 48

Note that repetitions can occur but 21 cannot be used twice.

B-12 Matrices

Given the matrices $A = \begin{bmatrix} 1 & -4 & 6 \\ 2 & -1 & 0 \\ 5 & -3 & 6 \end{bmatrix}$ and $B = \begin{bmatrix} 4 & 15 \\ -2 & 6 \\ 2 & 5 \end{bmatrix}$.

EXAMPLE 1: Find -2.5AB.

[MATRX] [►] [►] [1] <[A]> [3]

[ENTER] [3] [ENTER]

Choose matrix A and enter the dimensions.

[1] [ENTER] [(-)] [4] [ENTER]

[6] [ENTER] [2] [ENTER]

[(-)] [1] [ENTER] [0] [ENTER]

[5] [ENTER] [(-)] [3]

[ENTER] [6] [ENTER]

Enter the matrix elements.

Keystrokes	Description
MATRX ▶ ▶ 2 <[B]>	Choose matrix B and enter the dimensions.
3 ENTER 2 ENTER	
4 ENTER 15 ENTER	Enter the matrix elements.
(-) 2 ENTER 6 ENTER	
2 ENTER 5	
2nd QUIT	Return to the Home Screen. Note that [A] is the matrix A. This is different than A which represents the variable A.
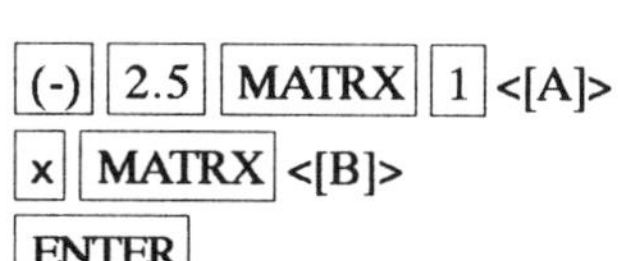	Enter the matrix operations. The result is $\begin{bmatrix} -60 & -52.5 \\ -25 & -60 \\ -95 & -217.5 \end{bmatrix}$.

EXAMPLE 2: Find A^{-1} using matrix A from Example 1.

Keystrokes	Description
	The operations are done from the Home Screen. The result rounded to three decimal places is: $A = \begin{bmatrix} -.167 & .167 & .167 \\ -.333 & -.667 & .333 \\ -.028 & -.472 & .194 \end{bmatrix}$
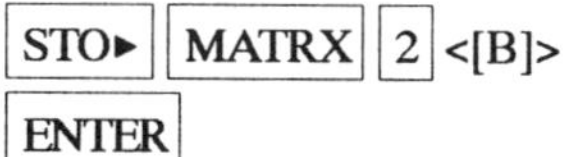	To inspect the result, store it as another matrix, say [B], recall it and and use the arrow keys to see the elements.

EXAMPLE 3: Solve the system of equations using the inverse of the coefficient matrix. Round answers to two decimal places.

$$\begin{aligned} x - 4y + 6z &= 2.5 \\ 2x - y &= 3.2 \\ 5x - 3y + 6z &= -3.4 \end{aligned}$$

Solution: The method of finding the inverse of a matrix was shown above in Section B-12 Example 2. Enter the constants as a 3x1 matrix B and find $A^{-1}B$.

Keystrokes	Description
MATRX ▶ ▶ 2 <[B]>	Enter the matrix dimensions.
3 ENTER 1 ENTER	
2.5 ENTER	Enter the matrix elements.
3.2 ENTER (-) 3.4	
2nd QUIT	Return to the Home Screen.
MATRX 1 <[A]> x^{-1}	Perform the matrix operations.
MATRX 2 <[B]> ENTER	The result is $\begin{bmatrix} -.45 \\ -4.10 \\ -2.24 \end{bmatrix}$. The solution to the system of equations to two decimal places is $x = -.45$, $y = -4.10$, $z = -2.24$.

EXAMPLE 4: Find the solution to the system of Example 3 above using Gauss-Jordan reduction. In other words, find the reduced form of the matrix $\begin{bmatrix} 1 & -4 & 6 & 2.5 \\ 2 & -1 & 0 & 3.2 \\ 5 & -3 & 6 & -3.4 \end{bmatrix}$.

Keystrokes	
MATRX ▶ ▶ 1 <[A]>	Name the matrix A. Enter the matrix dimensions.
3 ENTER 4 ENTER	
1 ENTER (-) 4 ENTER	Enter the matrix elements row by row.
6 ENTER 2.5 ENTER	
2 ENTER (-) 1 ENTER	
0 ENTER 3.2 ENTER	
5 ENTER (-) 3 ENTER	
6 ENTER (-) 3.4	
2nd QUIT	Return to the home screen.

After a calculation, some numbers may look like decimals. However upon storing, recalling and examining the matrix using the arrow keys you will find they are decimals that are very close to zero. This situation can be eliminated by setting the number of decimal places to a number, say 3 (see Section B-3 of this manual).

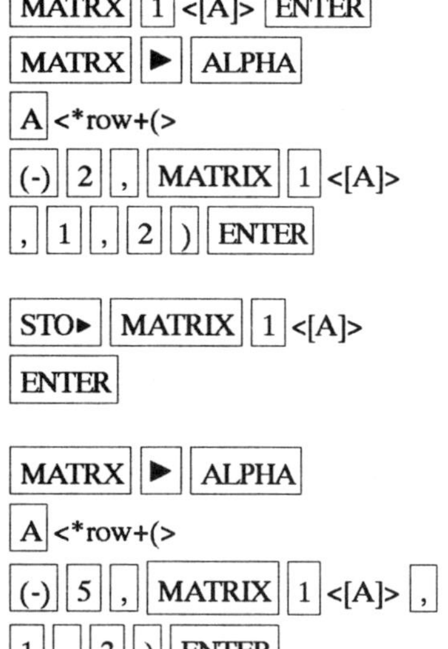

Display the matrix.

We wish to multiply by -2, matrix A, row 1 and add it to row 2. The result will be put in row 2.

```
*row+(-2,[A],1,2)
```

The result is $\begin{bmatrix} 1 & -4 & 6 & 2.5 \\ 0 & 7 & -12 & -1.8 \\ 5 & -3 & 6 & -3.4 \end{bmatrix}$.

Store the result as matrix A so you can perform the next operation.

Now multiply by -5, matrix A, row 1 and add it to row 3.

```
*row+(-5,[A],1,3)
```

The result is $\begin{bmatrix} 1 & -4 & 6 & 2.5 \\ 0 & 7 & -12 & -1.8 \\ 0 & 17 & -24 & -15.9 \end{bmatrix}$.

Store the result as matrix A.

STO▸ MATRIX 1 <[A]>
ENTER

Keystrokes	Explanation
[MATRX] [▶] [0] <*row(> [1] [÷] [7] [,] [MATRIX] [1] <[A]> [,] [2] [)] [ENTER]	Multiply by 1/7, matrix A, row 2. `*row(1÷7,[A],2)` The result is $\begin{bmatrix} 1 & -4 & 6 & 2.5 \\ 0 & 1 & -1.71 & -.26 \\ 0 & 17 & -24 & -15.9 \end{bmatrix}$ rounded to two decimal places.
[STO▸] [MATRIX] [1] <[A]> [ENTER]	Store the result as matrix A.

Continue in this manner until you get $\begin{bmatrix} 1 & 0 & 0 & -.45 \\ 0 & 1 & 0 & -4.10 \\ 0 & 0 & 1 & -2.24 \end{bmatrix}$. The solution to the system of equations is $x = -.45$, $y = -4.10$, $z = -2.24$.

> Note: To swap rows of a matrix use [8] <rowSwap(> from the [MATRX] MATH menu. The command rowSwap([A],2,3) will swap rows 2 and 3 in matrix A.
>
> Note: To add one row to another use [9] <row+(> from the [MATRX] MATH menu. The command row+([A],2,3) will add row 2 to row 3 in matrix A.

B-13 Logarithmic, Exponential, and Hyperbolic Functions

EXAMPLE 1: Graph $f(x) = 10^{3.2x}$ for $-.5 < x < .5$.

Solution: Set the WINDOW to [-.5,.5].1 by [-10,10]1. (See Appendix B Section B-7 Example 1 Method 1.)

Keystrokes	Explanation
[Y=] [10] [^] [(] [3.2] [X,T,θ] [)] [GRAPH]	Enter the functions and graph.

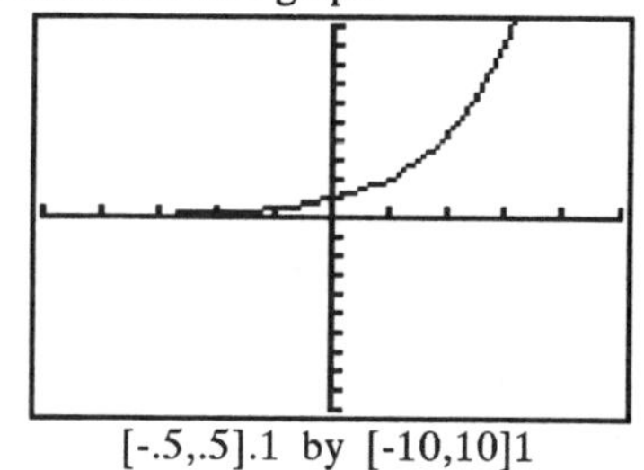

[-.5,.5].1 by [-10,10]1

***EXAMPLE** 2*: Graph $y = \dfrac{e^{x} + e^{-x}}{2}$.

Solution: This could be entered using [(] [2nd] [e^x] [X,T,θ] [+] [2nd] [e^x] [(-)] [X,T,θ] [)] [÷] [2]. However this is the hyperbolic cosine function. The following uses this built-in function. Recall that [e^x] represents the built-in function above the [LN] key.

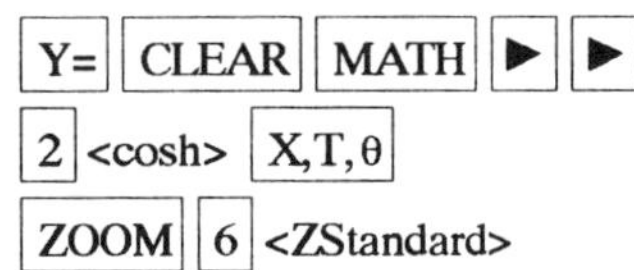

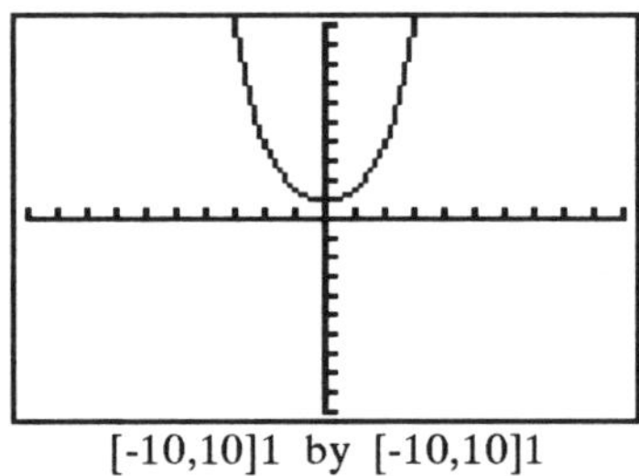

[-10,10]1 by [-10,10]1

EXAMPLE 3: Solve $\ln(x^2 + x) = 3x + 2$

Solution:

Y= CLEAR LN (X,T,θ ^ 2 + X,T,θ) ENTER — Enter the function $f(x) = \ln(x^2 + x)$ as Y1 in the function list.

CLEAR 3 X,T,θ + 2 ENTER — Enter the function $g(x) = 3x + 2$ as Y2 in the function list.

ZOOM 6 <ZStandard>

Graph them on the same graphing screen. Use ZOOM and TRACE to find the x value of the point of intersection. This will be the solution to the equation.

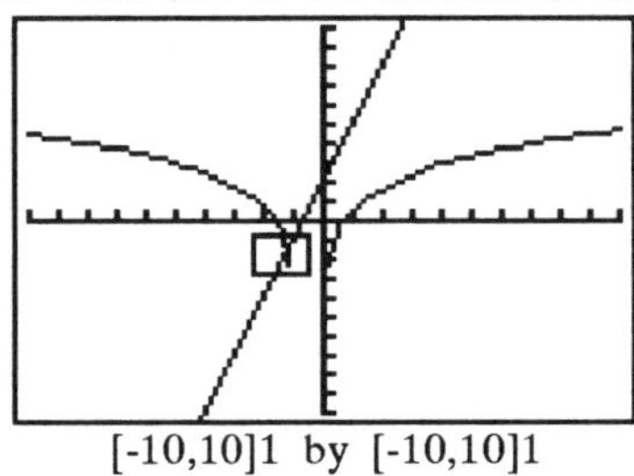

[-10,10]1 by [-10,10]1

The solution is -1.18, rounded to two decimal places.

B-14 Trigonometric Functions

EXAMPLE 1: Evaluate $f(x) = \cos x + \csc x$ at (A) 53°45'18" (B) 78.52 radians

Solution: (A)

Y= CLEAR COS X,T,θ + 1 ÷ SIN X,T,θ ENTER — Store the function.

MODE ▼ ▼ ▶ ENTER — Change the calculator to degree angle measure.

2nd QUIT

53 + 45 ÷ 60 + 18 ÷ 3600 STO▶ X,T,θ — Change the angle to decimal degrees while storing it as the variable X.

2nd Y-VARS 1 1 <Y1> ENTER — Evaluate the function. The result to two decimal places is 1.83.

Solution: (B)

Keystrokes	Explanation
[MODE] [▼] [▼] [ENTER]	Change the calculator to radian angle measure.
[2nd] [QUIT]	Since the function was stored in Part (A) above, we need only to store the angle measure and evaluate the stored function.
[78.52] [STO▸] [X,T,θ] [ENTER]	Store 78.52 as X.
[2nd] [Y-VARS] [1] [1] <Y1> [ENTER]	Evaluate the function. The result to two decimal places is 49.47.

EXAMPLE 2: Graph $f(x) = \cos x + \csc x$.

Solution: The function was entered in the calculator in Part (A) above.

Keystrokes	Explanation
[MODE] [▼] [▼] [▼] [▼] [▶] [ENTER] [2nd] [QUIT]	Change the graph mode to Dot. This will eliminate the "vertical lines."
[ZOOM] [7] <ZTrig>	This option automatically sets the viewing window variables to [-6.28,6.28]1.57 by [-3,3].25.

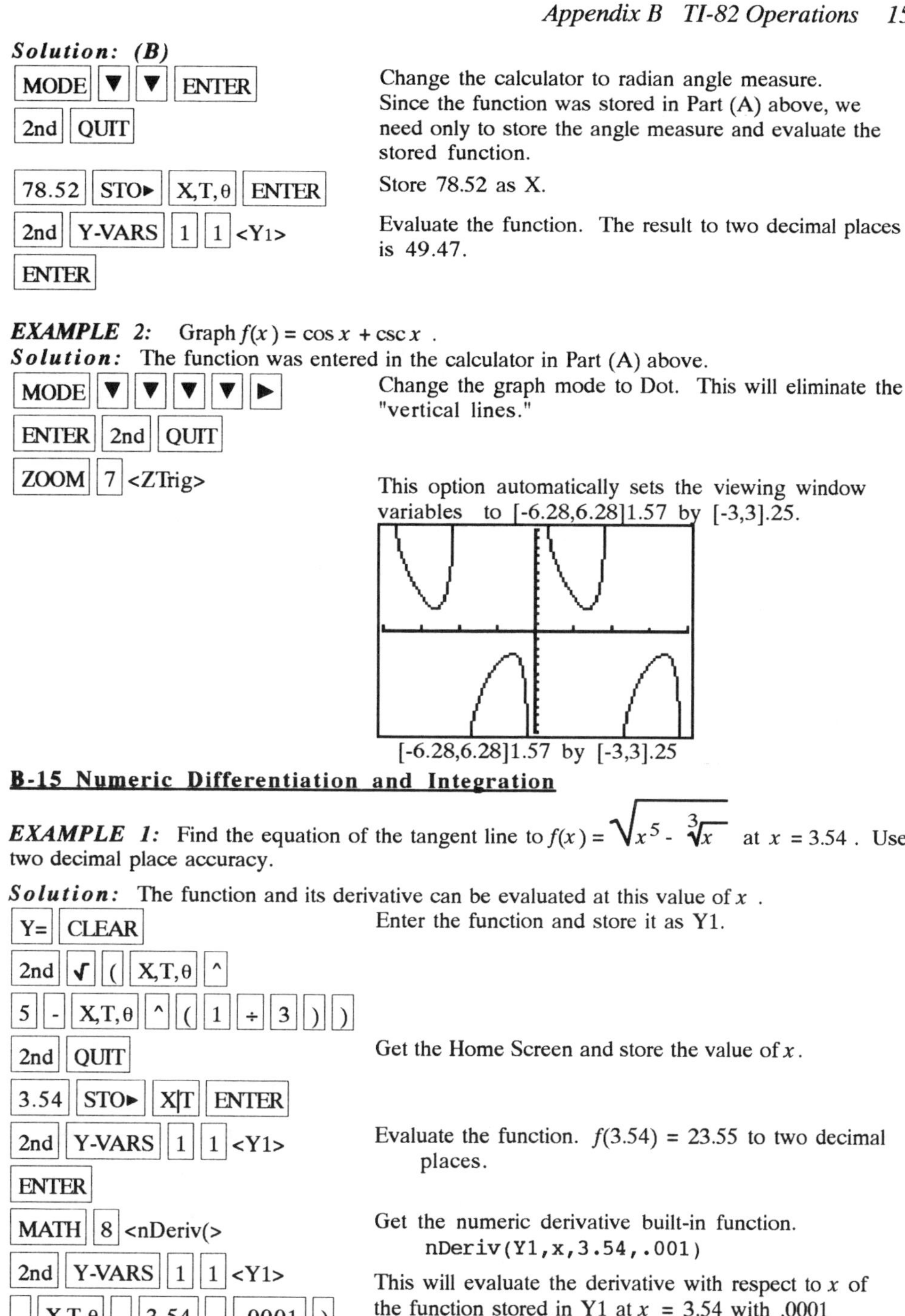

[-6.28,6.28]1.57 by [-3,3].25

B-15 Numeric Differentiation and Integration

EXAMPLE 1: Find the equation of the tangent line to $f(x) = \sqrt{x^5 - \sqrt[3]{x}}$ at $x = 3.54$. Use two decimal place accuracy.

Solution: The function and its derivative can be evaluated at this value of x .

Keystrokes	Explanation
[Y=] [CLEAR] [2nd] [√] [(] [X,T,θ] [^] [5] [-] [X,T,θ] [^] [(] [1] [÷] [3] [)] [)]	Enter the function and store it as Y1.
[2nd] [QUIT] [3.54] [STO▸] [X|T] [ENTER]	Get the Home Screen and store the value of x.
[2nd] [Y-VARS] [1] [1] <Y1> [ENTER]	Evaluate the function. $f(3.54) = 23.55$ to two decimal places.
[MATH] [8] <nDeriv(>	Get the numeric derivative built-in function. `nDeriv(Y1,x,3.54,.001)`
[2nd] [Y-VARS] [1] [1] <Y1> [,] [X,T,θ] [,] [3.54] [,] [.0001] [)]	This will evaluate the derivative with respect to x of the function stored in Y1 at $x = 3.54$ with .0001 tolerance. If the tolerance is not entered, the calculator will use .001.
[ENTER]	The result is 16.67.

The equation of the tangent line is $y - 23.55 = 16.67(x - 3.54)$.

EXAMPLE 2: Find $\int_{1.32}^{2.58} \sqrt{x^5 - \sqrt[3]{x}}\, dx$. Use two decimal place accuracy.

Solution:

Y= CLEAR 2nd √ (X,T,θ ^
5 - X,T,θ ^ (1 ÷ 3))

Enter the function and store it.

2nd QUIT

MATH 9 <fnInt(>

2nd Y-VARS 1 1 <Y1>

, X,T,θ , 1.32 , 2.58)

Get the numeric integration function built into the calculator function:

```
fnInt(Y1,X,1.32,2.58)
```

This will evaluate the definite integral . The result is 6.95 . The tolerance in calculating the definite integral is .00001.

B-16 Statistics

EXAMPLE 1: Given the following data find: (A) The mean. (B) The sample standard deviation . (C) An histogram with first class beginning at 0 and width 4. (D) Repeat Parts A and B after deleting the data value 89 from the list. How did eliminating this outlier effect the mean and standard deviation?

23	56	73	12	89	52	45	19	40	57	39
38	18	21	15	37	42	51	38	54	60	

Use one decimal place in answers.

Solution (A) and (B):

STAT 4 <ClrList>

2nd L1 ENTER

Clear the list named L1.

STAT 1 <Edit>

Get the data entry menu.

23 ENTER 38 ENTER
Etc.

Enter the data in the L1 list. Enter all data. There are 21 values.
If you make a mistake, use the arrow keys to highlight the number and type it in again.

To remove a data value, use the arrow keys to highlight number and press DEL .

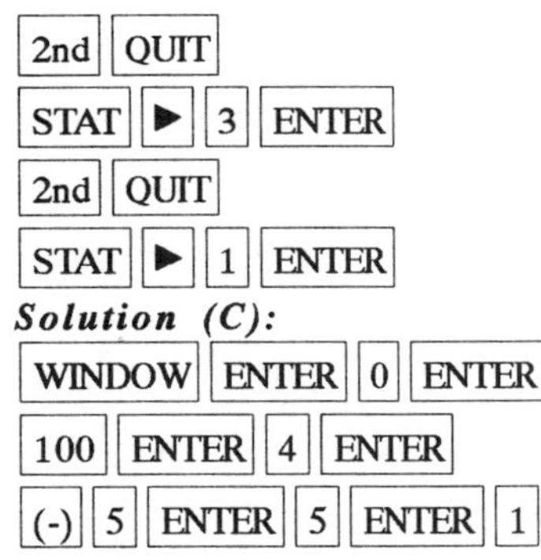

Exit the data entry screen.

Get the statistics menu again. Press 3 to see the setup. Under 1-Var Stats the Xlist should be L1 and the Freq should be 1. The mean is 41.86 and the sample standard deviation is 19.82.

The smallest value is 12 so let xMin=0.
The greatest value is 89. Let us set Xmax=100. Let Xscl=4 since this is the class width. Let Ymin=-5 to get some room below the histogram. We do not know the frequencies yet so let Ymax=5 for a first try. Let yScl=1.

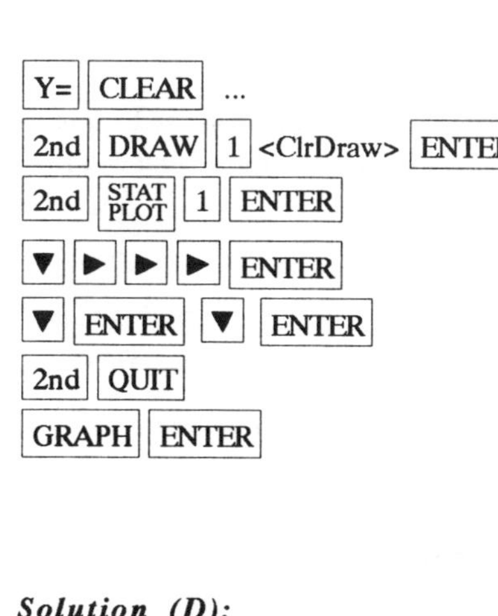

Clear all functions on the function list.

Clear all drawings from the display.

Set PLOT 1 to ON.
Select Type as Histogram.
Select Xlist as L1 and Freq as 1.
Graph the histogram.

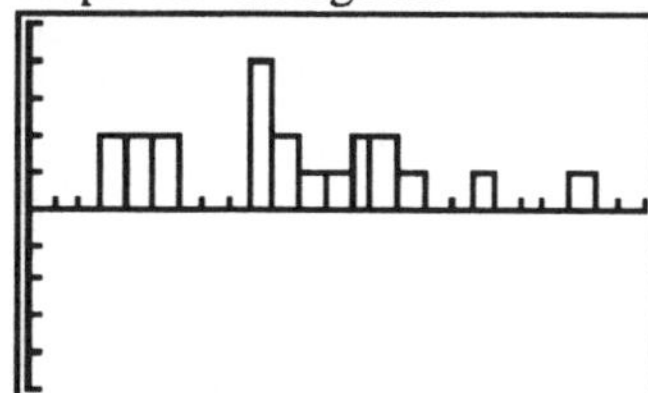

Solution (D):

Get the data list and arrow down to the 89. Press [DEL].

[STAT] [▶] [1] [ENTER]

Get the statistics menu and the one variable sample statistics. The mean is now 39.5 and the sample standard deviation is 17.05. Notice how much influence an outlier has!

EXAMPLE 2: Given the following bivariate data: (A) Find the correlation coefficient. (B) Find the least squares line. (C) Graph the least squares line and the scatter plot on the same graph. (D) Plot the scatter diagram without the least squares line. (E) Sort the data on x and join the points using straight line segments.

x	5	8	10	29	9	24
y	3	4	8	20	3	19

Solution (A):

[STAT] [4] <ClrList> [2nd] [L1] [,] [2nd] [ENTER]

Clear list L1 and list L2.

[STAT] [1] <Edit>

Get the lists.

[5] [ENTER] [8] [ENTER] [10] [ENTER] [29] [ENTER] [9] [ENTER] [24] [ENTER]

Enter the data in list L1.

[▶] [3] [ENTER] [4] [ENTER] [8] [ENTER] [20] [ENTER] [3] [ENTER] [19] [ENTER]

Move to list L2. Enter the data in list L2. The data for y in L2 must correspond pairwise to the data for x in L1.

[STAT] [▶] <Calc> [3] <Set Up>
[▼] [▼] [ENTER] [▼] [▶] [ENTER]

Set up the calculator to use L1 as the x variable and L2 as the y variable in the calculations.

[STAT] [▶] <Calc> [9] <LinReg(a+bx)> [ENTER]

This give you the linear correlation coefficient, the constant term and the coefficient on x term for the least squares line. $r = .98$.

Solution (B):

Keystrokes	Explanation
VARS 5 <Statistics> ► ► 7 <RegEQ>	The least squares line equation is displayed. It is $\hat{y}$ = -1.72 + .79x with the coefficients rounded to two decimal places.

Solution (C):

Keystrokes	Explanation
WINDOW ENTER 0 ENTER 30 ENTER 1 ENTER (-) 5 ENTER 30 ENTER 1	Set the graph WINDOW. Observe the data to determine the WINDOW values. Set the y value lower than needed so if you trace, the coordinates at the bottom of the screen do not cover up the graph.
Y= CLEAR ...	Clear all functions on the function list.
2nd DRAW 1 <ClrDraw> ENTER	Clear all drawings from the display.
	Turn the other STAT PLOTS off.
Y= VARS 5 ► ► 7 <RegEQ>	The least squares line equation will be put into the function list so you can graph it.
2nd STAT PLOT 1 <Plot 1> ENTER ▼ ENTER ▼ ENTER ▼ ► ENTER ▼ ENTER GRAPH	Set up the calculator to get the scatter plot with the line.

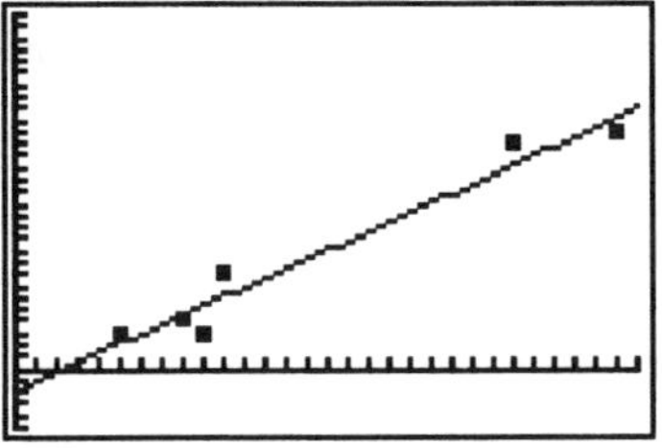

Solution (D):

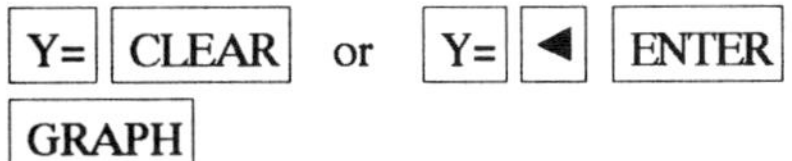

GRAPH

Clear the functions from the list or deselect them so they won't graph.
Plot the points on the graph.

Solution (E):

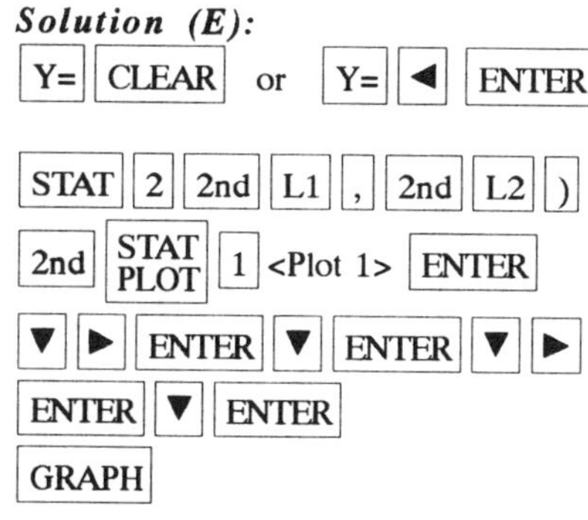

Clear the functions from the list or deselect them so they won't graph.

Sort the data on the x variable. The list L1 will be sorted in ascending order. The list L2 will sort along with L1 as pairs.
Set the calculator to draw line segments joining the points.
This will join the points together with straight line segments in the order they appear in the data list. This is why we sorted the data on x first.

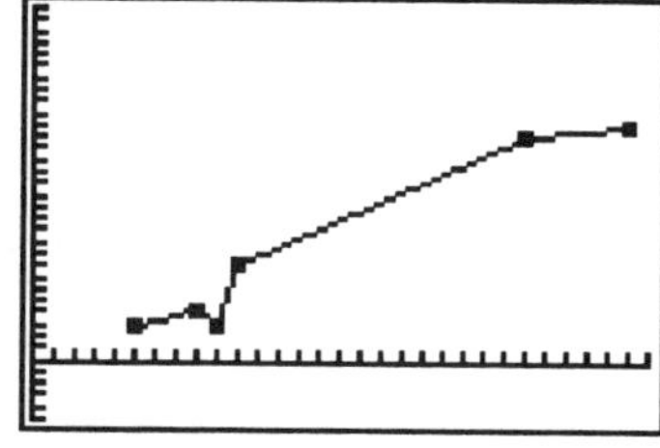

B-17 Setting the Display Mode for Numbers

Numbers can be entered into the calculator in scientific notation, a set number of significant digits, or a fixed number of decimal places.

EXAMPLE 1: Calculate $(9.35417\times 10^{4})(-3.64389\times10^{-6})$ using:
(A) scientific notation (B) two significant digits (C) four decimal place accuracy.

Solution (A):

MODE ▶ ENTER 2nd QUIT — Set the calculator to scientific mode with floating decimal point.

9.35417 2nd EE 4 × (-)
3.64389 2nd EE (-) 6
ENTER

The result displayed on the calculator is -3.408556652E-1. This means $-3.408556652\times10^{-1}$.

Solution (B):

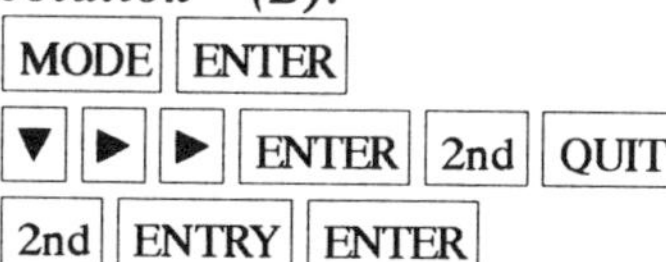

Set the calculator to scientific notation with 1 fixed decimal point instead of a floating decimal. This will give two significant digits.
Return to the Home Screen.
Use the replay key to replay the last calculation done in Solution (A). The result to two significant digits is -3.4E-1 .

Solution (C):

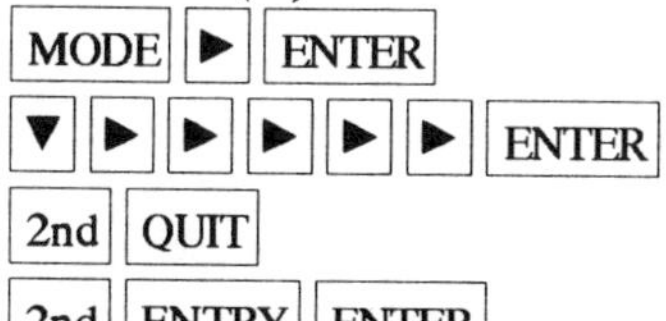

Set the calculator to normal mode with 4 digits.

Return to the Home Screen. Use the replay key to replay the last calculation.
The result to four decimal places is -.3409.

You may now wish to return your calculator to its default settings (all items on the mode screen having the left item highlighted). The examples throughout this manual are worked using the default settings unless otherwise noted.

NOTES

Appendix C

TI-85 Operations

C-1 Operations

Press ON to turn on the calculator.

Press 2nd OFF to turn off the calculator.

Press CLEAR to clear the screen.

Press EXIT to clear any menus displayed at the bottom of the screen.

Press 2nd ▲ to make the display darker.

Press 2nd ▼ to make the display lighter.

Press 2nd + to get the RESET screen .

Press F3 <RESET> and then F1 <ALL> to clear all the memories.

RAM	DELET	RESET		
ALL	MEM	DELTS		

Press F4 <YES> to clear.

Home Screen

The screen where calculations are done is called the Home Screen. You can always get to the Home Screen by pressing 2nd QUIT. Note: QUIT is above the EXIT key.

> The item above the key will be referenced as if it is on the key. For example, QUIT, which is above the EXIT key, will be referenced from now on as 2nd QUIT.

2nd

This key must be used to access the calculator functions above and to the left of a key cap. A new cursor is displayed, ↑. These functions are in yellow on the calculator face.

ALPHA

This key is used to access the calculator functions and memory locations above and to the right of a key cap. These function are in light blue on the calculator face. The cursor changes to an A when this key is pressed.

MODE

This is found above the MORE key and is used to set the operating mode of the calculator. When the highlighted items are on the left, the calculator is in default mode. To choose a different mode: Use the arrow keys to choose the item you want to change to, then press the [ENTER] key. The new selection will now be highlighted. Press [2nd] [QUIT] to return to the Home Screen.

<u>Menus</u>

Menus are listed across the bottom of the screen. To choose a menu item press the function key immediately below the menu item. In this manual, these functions will be referred to as the key to press in a box followed by the function in angle brackets. For example, the trace function of the calculator is found on the [GRAPH] menu above the [F4] key is denoted by [F4] <TRACE>.

<u>Error Correction</u>

Use the [2nd] [INS] key to insert characters. ([INS] is above the [DEL] key.)

Use the [DEL] key to delete characters.

Use the arrow keys [◄] or [►] to move to the left or right.

Use the up arrow key [▲] to move up one line.

Use the down arrow key [▼] to move down one line.

[2nd] [ENTRY] is used to replay the last executed operation. ([ENTRY] is above the [ENTER] key.)

C-2 Calculating

Many calculator functions are found above key caps or in menus. For example the square root function is above the x^2 key but the absolute value is found in the [MATH] menu.

EXAMPLE 1: Calculate $|4^2 - \sqrt{34}\,|$

Solution:

[2nd] [MATH] [F1] <NUM> [F5] <abs> [(] [4] [x^2] [-] [2nd] [√] [34] [)] [ENTER]. [ENTER]

tells the calculator to calculate. The result displayed is 10.1690481052.

C-3 Plotting Points and Drawing Lines

EXAMPLE 1: (A) Plot the points (-3, 6) and (1,5). (B) Draw the line segment having these points as the endpoints.

Solution (A):

[GRAPH] [F3] <ZOOM> [F4] <ZSTD>

We first need to set the viewing window dimensions. [ZOOM] [F4] <Standard> automatically sets the dimensions to Xmin=-10, Xmax=10, Xscl=1, Ymin=-10, Ymax=10, Yscl=1 and Xres=1. This will be denoted as [-10, 10]1 by [-10, 10]1 in this manual.

[EXIT] [EXIT]

[GRAPH] [MORE] [F2] <DRAW>

[MORE] [F5] <CLDRW>

[2nd] [QUIT]

Get the [DRAW] menu and clear the drawings from the calculator screen.

[2nd] [CATALOG] [▼] ... [▼]
[ENTER]

Get the [CATALOG] listing all of the functions. Use the down arrow until the pointer points to PtOn(. Press [ENTER].

[(-)] [3] [,] [6] [)] [ENTER]

Enter the coordinates of the first point.

[2nd] [ENTRY] [◄] ... [◄]
[1] [,] [5] [)] [ENTER]

Use the replay feature of the calculator to get the last entry. Use the left arrow and change the coordinates to those of the second point. Delete any unwanted characters using [DEL].

Solution (B):

[2nd] [QUIT]

Return to the Home Screen.

[2nd] [CATALOG] [▼] ... [▼]
[ENTER]

Get the [CATALOG] listing all of the functions. Use the down arrow or [F1] <PAGE↓> until the pointer points to Line(. Press [ENTER].

[(-)] [3] [,] [6] [,]
[1] [,] [5] [)] [ENTER]

Enter the coordinates of the first point and the second point separated by commas. Press enter to draw the line segment.

C-4 Storing and Evaluating an Expression

EXAMPLE 1: Store the expression $\frac{-b+\sqrt{b^2-4ac}}{2a}$ and evaluate it to find one solution of $3x^2+4x-5=0$.

Solution:

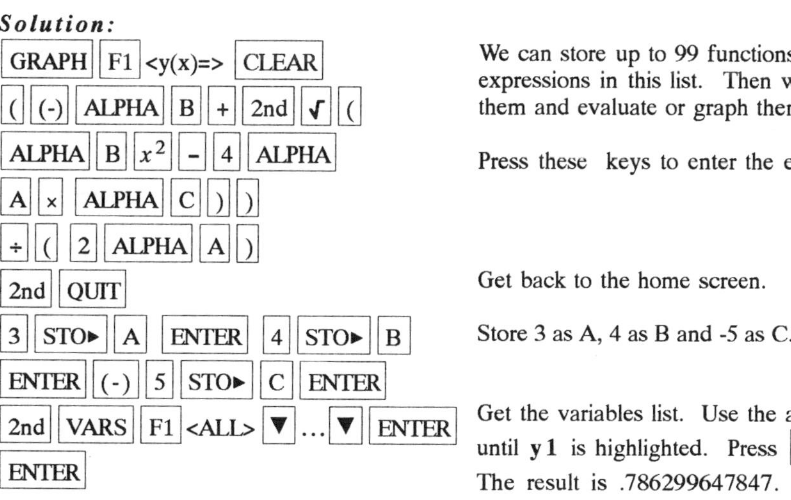

[GRAPH] [F1] <y(x)=> [CLEAR]

We can store up to 99 functions or algebraic expressions in this list. Then we can recall them and evaluate or graph them.

[(] [(-)] [ALPHA] [B] [+] [2nd] [√] [(]
[ALPHA] [B] [x^2] [-] [4] [ALPHA]
[A] [×] [ALPHA] [C] [)] [)]
[÷] [(] [2] [ALPHA] [A] [)]

Press these keys to enter the expression.

[2nd] [QUIT]

Get back to the home screen.

[3] [STO▸] [A] [ENTER] [4] [STO▸] [B]
[ENTER] [(-)] [5] [STO▸] [C] [ENTER]

Store 3 as A, 4 as B and -5 as C.

[2nd] [VARS] [F1] <ALL> [▼] ... [▼] [ENTER]
[ENTER]

Get the variables list. Use the arrow keys until **y1** is highlighted. Press [ENTER]. The result is .786299647847.

EXAMPLE 2: Given $f(x) = \sqrt[3]{x^4+2x}$ and $g(x) = 4^{2x-90}$. Find $f(x) + 3\sqrt{g(x)}$ at $x = 45.2$.

Solution:

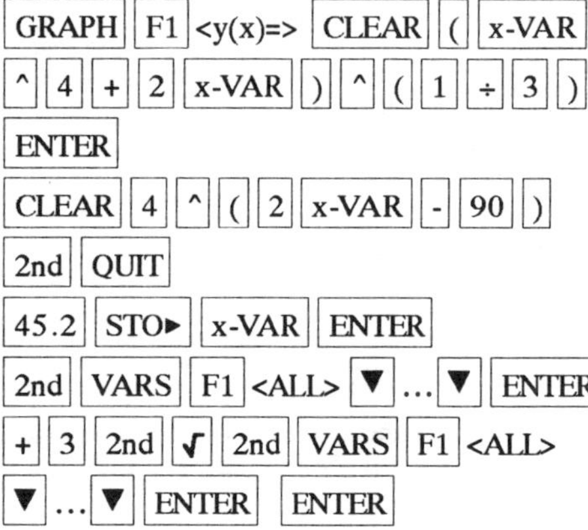

Enter the functions as y1 and y2. Press CLEAR to delete the old y1 function. Note that lower case x is used in functions. Upper case and lower case letters are different. Values can be stored for each of these.

Store 45.2 as x.

Form $f(x) + 3\sqrt{g(x)}$ as y1 + 3√ y2 getting y1 and y2 from the variables list. The expression will be evaluated when you press ENTER.

The result is 164.97 to two decimal places.

C-5 Evaluating Inequalities

EXAMPLE 1: Given the inequality $x^2 - 3t < 5x + 9t^2$. Test this inequality for $x = 1.89$ and $t = -4.52$.

Solution:

1.89 STO▸ 2nd alpha x-VAR
ENTER (-) 4.52 STO▸ 2nd alpha

Store the value of the variables. Lower case letters are used.

T ENTER x-VAR ^ 2 – 3
2nd alpha T 2nd TEST F2 :<< > 5
x-VAR + 9 2nd
alpha T ^ 2 ENTER

Enter the expression and evaluate.

The result is 1. This means the inequality is true for these values of the variables. If the inequality was false, the result would have been 0.

C-6 Graphing

EXAMPLE 1: Graph $y = 3x^2-6$ and $y=3x^2+1$ on the same graphing screen.

Solution:

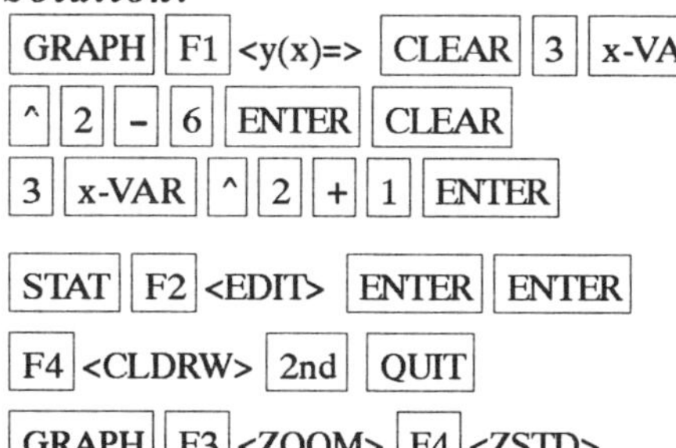

GRAPH F1 <y(x)=> CLEAR 3 x-VAR ^ 2 - 6 ENTER CLEAR 3 x-VAR ^ 2 + 1 ENTER

Store the functions. Remember to clear out the old function whenever entering a new one. Clear or deselect all others by placing the cursor at the expression for the function and pressing F5. Now this function will not graph. Note that the = sign is no longer highlighted.

STAT F2 <EDIT> ENTER ENTER F4 <CLDRW> 2nd QUIT

Clear drawings.

GRAPH F3 <ZOOM> F4 <ZSTD>

Exit the function menu. Press F3 and then F4 to get the standard viewing window. This will automatically set the graphing screen to have $-10 < x < 10$ and $-10 < y < 10$ and graph the functions.

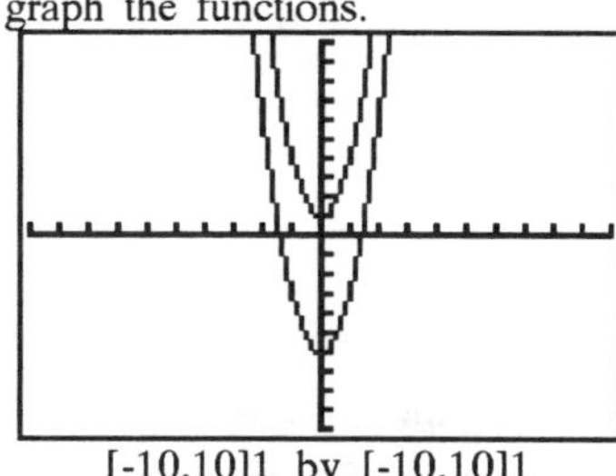

[-10,10]1 by [-10,10]1

CLEAR 2nd QUIT

Note that pressing CLEAR will remove the menus from the bottom of the screen.

ZSTD automatically sets the graphing screen limits to $-10<x<10$ and $-10<y<10$. These graph screen limits will be denoted in this manual by the notation [-10,10]1 by [-10,10]1. The graph screen limits can be set by pressing F2 <RANGE> on the GRAPH menu and entering the desired values.

F2 <RANGE> means to choose the RANGE option from the menu on the screen above the F2 key.

The function will <u>not</u> graph if you choose a function on the F1 <y(x)=> menu and then press F5 <SELCT>. You will notice that the highlighting disappears from the = sign. You can still evaluate this expression for stored values of the variables. To select the function, press F5 <SELCT> again.

EXAMPLE 2: Graph $f(x) = \begin{cases} .5x + 1 & x \le 2 \\ x^2 + 2 & x > 2 \end{cases}$

Solution:

The calculator will only graph functions for values of x for which the function is defined. So we write this piecewise-defined function in a form where the pieces are defined only for the values of x that we want. Enter two functions as $y1=(.5x+1)\div(x\le2)$ and $y2=(x^2+2)\div(x>2)$. The first function is defined with $x \le 2$ and the second function is defined when $x > 2$.

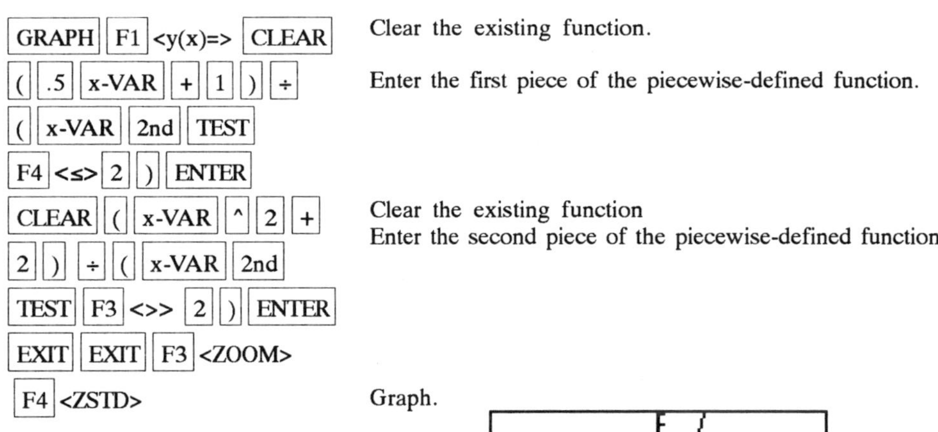

Keystrokes	
GRAPH F1 <y(x)=> CLEAR	Clear the existing function.
(.5 x-VAR + 1) ÷ (x-VAR 2nd TEST F4 <≤> 2) ENTER	Enter the first piece of the piecewise-defined function.
CLEAR (x-VAR ^ 2 + 2) ÷ (x-VAR 2nd TEST F3 <>> 2) ENTER	Clear the existing function Enter the second piece of the piecewise-defined function.
EXIT EXIT F3 <ZOOM> F4 <ZSTD>	Graph.

[-10,10]1 by [-10,10]1

C-7 Trace, Zoom and Range

Graph $y = 3x^2-6$ (see Example 1 from Section C-6).

F4 <TRACE>	You see a new cursor near the center of the graph screen and the coordinates of that point on the graph displayed at the bottom of the screen.
◀ or ▶	Use right and left arrow keys and move the cursor along the graph. If we had the graph of two or more functions displayed, we could use the up and down arrow keys to move between the functions.
▲ or ▼	

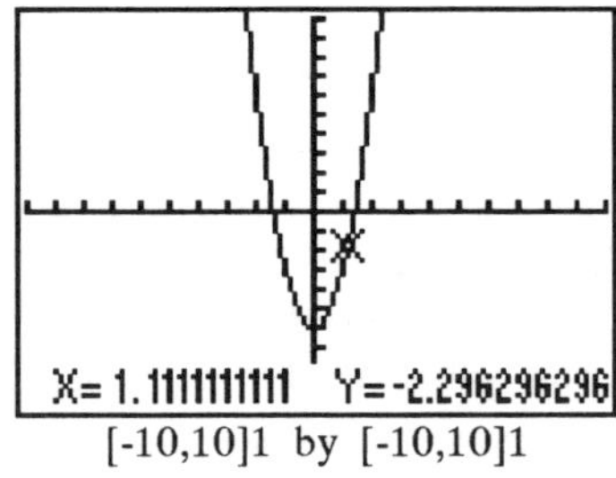

[-10,10]1 by [-10,10]1

We can zoom in on a point to get a closer look. There are three ways to do this:

1. Change the RANGE values.
2. Set the ZOOM FACTORS and ZOOM IN.
3. Use the ZOOM BOX feature of the calculator.

EXAMPLE 1: Find the x intercept for $x>0$ for $y = 3x^2-6$. Use three decimal place accuracy.

METHOD 1 Set the RANGE

Using trace on the functions graphed above we see the x intercept is near 1.5. So we shall set the RANGE closer to that value.

[GRAPH] [F2] <RANGE> [1]
[ENTER] [2] [ENTER] [.1] [ENTER]
[(-)] [.5] [ENTER] [.5] [ENTER] [.1]
[GRAPH] [F4] <TRACE>

Set the RANGE to [1,2].1 by [-.5,.5].1.

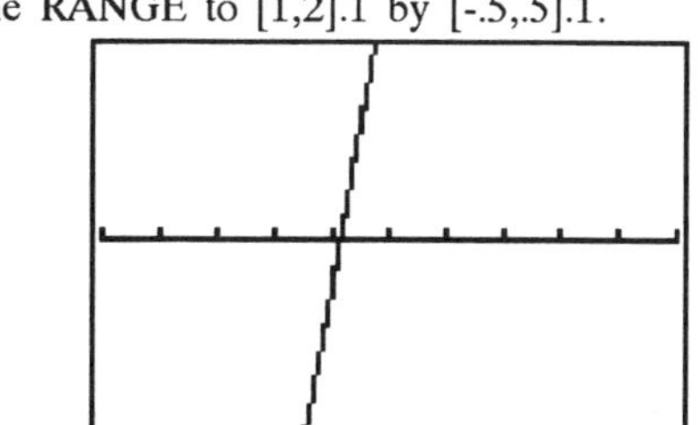

[1,2].1 by [-.5,.5].1

Press [EXIT] to restore the menu at the bottom of the screen.

Use TRACE again to get a better approximation. The trace cursor always starts at the middle x value of the RANGE. It may not be on the display. Watch the values for x at the bottom of the screen.

We see that the x intercept is between 1.41 and 1.42. Set the RANGE to [1.41,1.42].01 by [-.01,.01].001 and graph again. We now find the x intercept to 1.414 accurate to three decimal places.

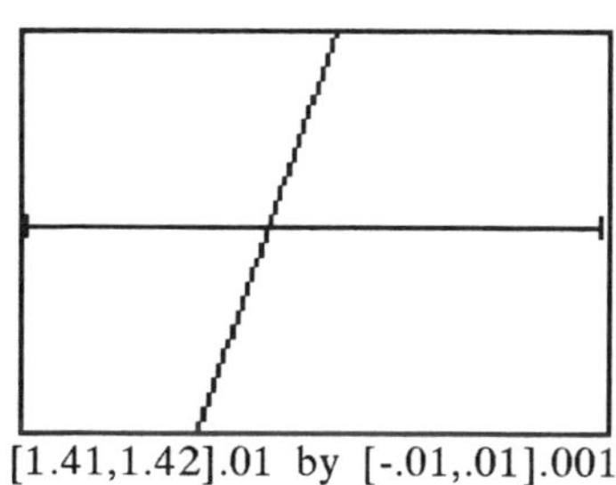

[1.41,1.42].01 by [-.01,.01].001

METHOD 2 Zoom In

[GRAPH] [F3] <ZOOM>

[F4] <ZSTD>

Get the [-10,10]1 by [-10,10]1 display as shown at the beginning of this section.

[MORE] [MORE] [F1] <ZFACT>

[5] [ENTER] [5]

Set the zoom factors to 5 and 5. The graph will be 5 times as large in both the x and y directions.

[ENTER] [F3] <ZOOM> [F2] <ZIN>

[◀] or [▶] or [▲] or [▼]

Use the [◀] or [▶] arrow keys to get the cursor to where you want the center of the new display, say at (1.42, 0).

[EXIT]

Press [EXIT] to restore the menu at the bottom of the screen.

ENTER — The new display will be centered where you placed the cursor.

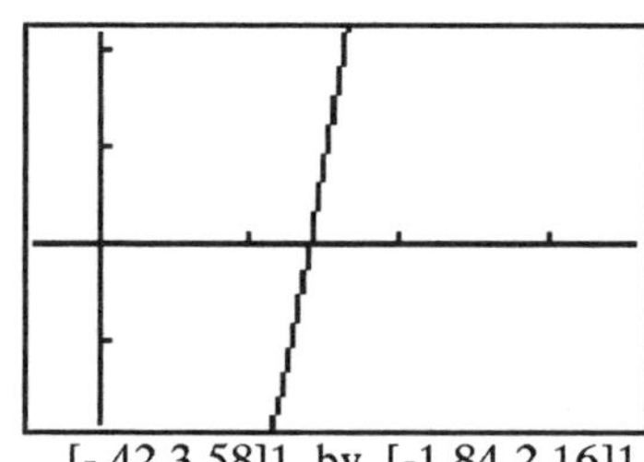

[-.42,3.58]1 by [-1.84,2.16]1

Now use TRACE again to get a new approximation to the x intercept.
Repeat the Zoom In process as many times as necessary to get accuracy to three decimal places.
The x intercept is 1.414 .

<u>METHOD 3 Zoom Box</u>

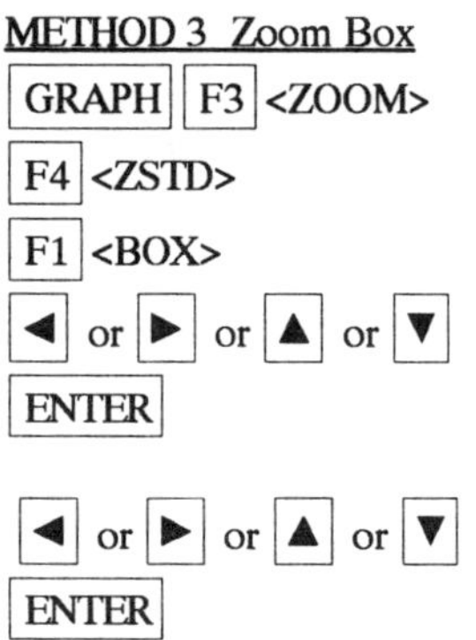

GRAPH F3 <ZOOM>
F4 <ZSTD> — Get the [-10,10]1 by [-10,10]1 display as shown at the beginning of this section.

F1 <BOX> — Get the zoom box option.

◄ or ► or ▲ or ▼
ENTER — Use the arrow keys repeatedly to get the cursor close to but a little above and to the left of the x intercept, say at X=1.1111111111 Y=.9677419355 Press ENTER to place the upper corner of the zoom box at this point.

◄ or ► or ▲ or ▼
ENTER — Now move the cursor down and to the right of the x intercept, say at X=2.0634920635 Y=-.6451612903. Press ENTER again. The graph will be drawn again with the box as the outside of the new display.

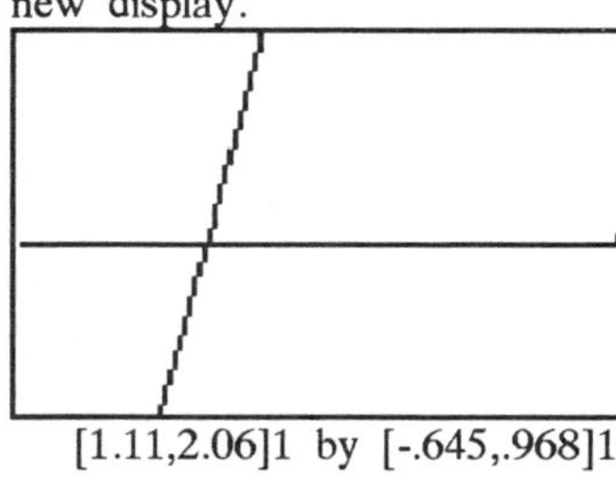

[1.11,2.06]1 by [-.645,.968]1

Use trace again to get another estimate of the x intercept. Use Zoom Box a second time to get a better approximation of the x intercept. The x intercept is 1.414 .

<u>C-8 Choosing the RANGE</u>

There is no one right way to determine the RANGE that should be used for the graph of a function. Three methods to try are:

1. Graph using the standard screen and then zoom out.
2. Analyze the features of the function and then choose a RANGE.
3. Evaluate the function at several values.

EXAMPLE 1 Graph $y = .03x^3 + .02x^2 - 5x + 10$.

Solution:

METHOD 1 Graph using the standard screen and then zoom out

Enter the function as .03 X ^ 3 + .02 X ^ 2 - 5 X + 10.

Press EXIT F3 <ZOOM> F4 <ZSTD> to get the [-10,10]1 by [-10,10]1 display.

Press F3 <ZOOM> MORE MORE F1 <ZFACT> and set the zoom factors to xFact=4 and yFact=4. Press F3 <ZOOM> F3 <ZOUT>.

Move the cursor so it is where you want the center of the new screen to be, and press ENTER. The result is shown on the next page. The screen has been zoomed out to give us a better view of the graph. Note the double lines as the axes. This is because the Xscl and Yscl values are too close together. They do not change when zoom is used.

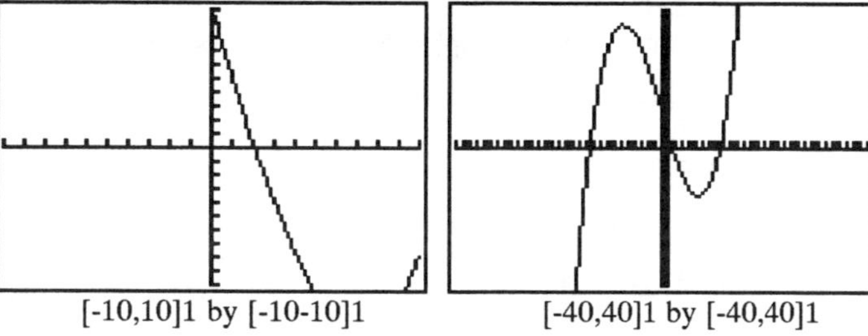

[-10,10]1 by [-10-10]1 [-40,40]1 by [-40,40]1

METHOD 2 Analyze the features of the function and then choose a RANGE

We will analyze the term with highest power on x since this is the dominant term of the function. We see that the leading coefficient is .03. This means that the function will increase 1 unit every time x^3 increases since $\frac{1}{.03} \approx 30$ units (x increases $\sqrt[3]{30} \approx 3$ units). A good first choice for the limits on the x axis is Xmax = 10×(unit increase in x)=30 and Xmin=-30. A good number of scale marks for the axes is 20. Hence the Xscl could be set at $\frac{\text{Xmax-Xmin}}{20} = \frac{30-(-30)}{20} = 3$. The constant term is 10. This means the y intercept is 10. Hence a good first choice for the y axis limits is Ymax=10 and Ymin=-10. The Yscl is then $\frac{\text{Ymax-Ymin}}{20} = \frac{10-(-10)}{20} = 1$. Set the RANGE at [-30,30]3 by [-10,10]1. As can be seen, the RANGE needs to be adjusted further to get a good display.

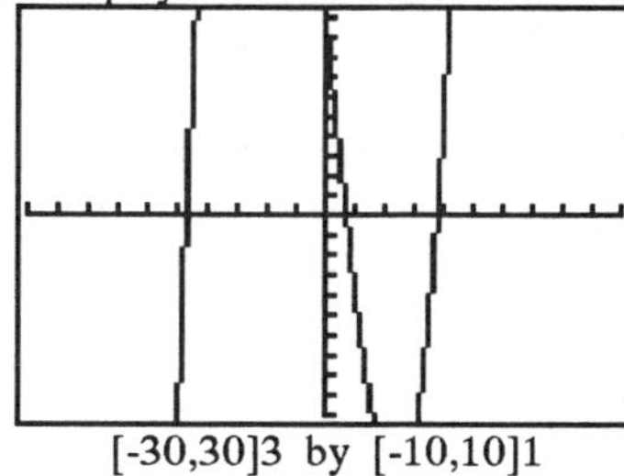

[-30,30]3 by [-10,10]1

METHOD 3 Evaluate the function at several values of x and then choose a RANGE. The choice of x values is arbitrary. Store the value of x in memory location x , then evaluate the function. Repeat for all values of x.

x	$f(x)$
-25	-321.25
-10	32.00
0	10.00
10	-8.00
25	366.25

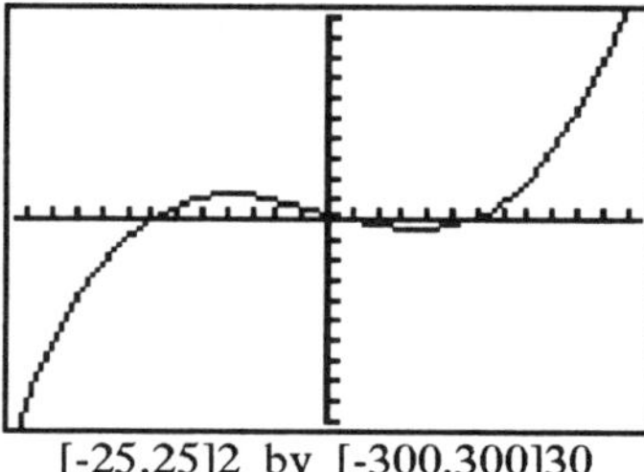
[-25,25]2 by [-300,300]30

Choose the Xscl and Yscl so there are about 20 marks on each axis (see METHOD 1 above).

So a good first choice for the RANGE could be [-25,25]2 by [-300,300]30.

C-9 Equations in One Variable

An equation in one variable can be solved graphically in two ways:

1. Algebraically move all terms to one side of the equation. Create a function $f(x)$=(the expression). Graph the function. Use zoom and trace to find the x intercepts. The set of x intercepts is the solution to the equation.
2. Create two functions: $f(x)$ = (the left side of the equation) and $g(x)$ = (the right side of the equation). Graph both functions. Use zoom and trace to find the x value of the points of intersection of the two graphs. The set of x values is the solution to the equation.

EXAMPLE: Solve $\sqrt[3]{x+2} = 4x^2 - 3\sqrt{x+4}$. Use two decimal places.
Solution:

METHOD 1 Write the equation as $\sqrt[3]{x+2} - (4x^2 - 3\sqrt{x+4}) = 0$. Create the function

$f(x) = \sqrt[3]{x+2} - (4x^2 - 3\sqrt{x+4})$. Enter this function into the calculator and graph.

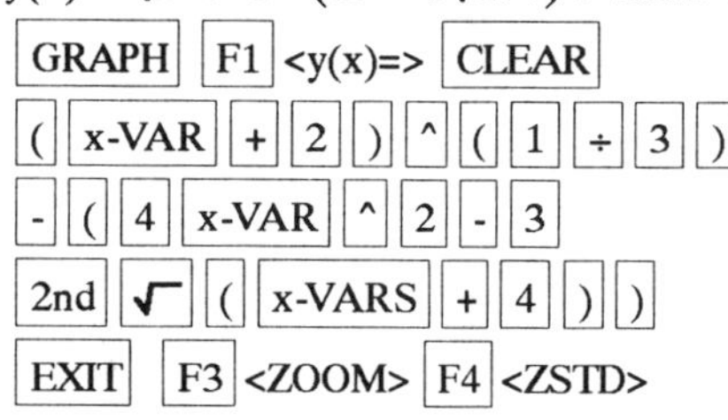

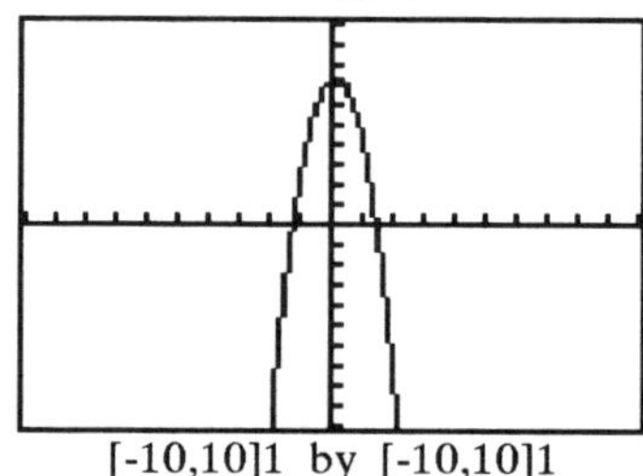
[-10,10]1 by [-10,10]1

We see the function crosses the x axis in two places. Use zoom and trace to find these x values. The solution to this equation is x = -1.22 and x = 1.46.

METHOD 2 Create the two functions $f(x) = \sqrt[3]{x+2}$ and $g(x) = 4x^2 - 3\sqrt{x+4}$. Graph these and find the x values of the intersection points.

[GRAPH] [F1] <y(x)=> [CLEAR] [(] [x-VAR]
[+] [2] [)] [^] [(] [1] [÷] [3] [)]
[ENTER] [4] [x-VAR] [^] [2]
[-] [3] [2nd] [√] [(] [x-VAR] [+] [4] [)]
[EXIT] [F3] <ZOOM> [F4] <ZSTD>

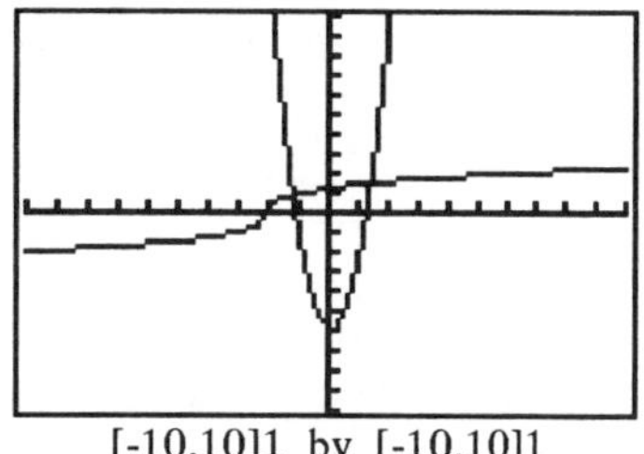

[-10,10]1 by [-10,10]1

We see the functions intersect in two places. Use zoom and trace to find the coordinates of these points. The intersection points are at (-1.22 .92) and (1.46, 1.51). So the solution to this equation is $x = -1.22$ and $x = 1.46$.

C-10 Inequalities in One Variable

An inequality in one variable can be solved graphically in two ways:

1. Algebraically move all terms to one side of the inequality and change the sign to an equal sign. Create a function $f(x)$=(the expression). Graph the function. Use zoom and trace to find the x intercepts. The set of x intercepts are the cutoff values for the solution. Determine the intervals over which the inequality is true and write the solution.
2. Create two functions: $f(x)$ = (the left side of the equation) and $g(x)$ = (the right side of the equation). Graph both functions. Use zoom and trace to find the x value of the points of intersection of the two graphs. The set of x coordinates are the cutoff values for the solution. Determine the intervals over which the inequality is true and write the solution.

EXAMPLE: Solve $\sqrt[3]{x+2} < 4x^2 - 3\sqrt{x+4}$. Use two decimal places

Solution:

METHOD 1 Write the inequality as: $\sqrt[3]{x+2} - (4x^2 - 3\sqrt{x+4}) < 0$. Change the sign to an equal sign: $\sqrt[3]{x+2} - (4x^2 - 3\sqrt{x+4}) = 0$. Create the function $f(x) = \sqrt[3]{x+2} - (4x^2 - 3\sqrt{x+4})$, graph, and find the x intercepts. This was done above in Appendix Section C-9 Method 1 of this manual. The cutoff values are $x = -1.22$ and $x = 1.46$. Observing the graph we see it is below the x axis for $x < -1.22$ and $x > 1.46$. Hence the solution to this inequality can be written as $(-\infty,-1.22)\cup(1.46,\infty)$.

METHOD 2 Create the two functions $f(x) = \sqrt[3]{x+2}$ and $g(x) = 4x^2 - 3\sqrt{x+4}$. Graph these and find the x values of the intersection points. This was done above in Appendix Section C-9 Method 2 of this manual. The intersection points are at (-1.22, .92) and (1.46, 1.51). Inspecting the graph we see that $f(x) < g(x)$ for $x < -1.22$ and $x > 1.46$. Hence the solution to this inequality can be written as $(-\infty,-1.22)\cup(1.46,\infty)$.

C-11 Permutations and Combinations

EXAMPLE 1: Find: (A) $P_{19,13}$ (B) $C_{19,13}$

Solution (A) : This is a direct calculation.

[19] [2nd] [MATH] [F2] <PROB>
[F2] <nPr> [13] [ENTER]

Enter the numbers and get the built-in function from the MATH menu. The result is displayed in scientific notation as 1.68951528346E14.

Solution (B) : This is a direct calculation.

Keystrokes	Explanation
[19] [2nd] [MATH] [F2] <PROB> [F3] <nCr> [13] [ENTER]	Enter the numbers and get the built-in function from the MATH menu. The result is 27132.
[EXIT] [EXIT]	To return to the home screen.

EXAMPLE 2: Generate ten random numbers at least 0 and less than 50.
Solution:

Keystrokes	Explanation
[5] [STO▸] [2nd] [MATH] [F2] <PROB> [F4] <rand>) [ENTER]	We need to store a "seed" number in the storage location RAND. This tells the random number generator where to start.
[EXIT] [F1] <NUM> [F2] <IPart> [(] [50] [x] [EXIT] [F2] <PROB> [F4] <rand> [)] [ENTER] [ENTER]	RAND generates a number greater than 0 and less than 1. If we multiply by 50 and take the integer part, the result is a number greater than or equal to 0 and less than 50.
[ENTER] [ENTER] [ENTER] [ENTER] [ENTER] [ENTER] [ENTER] [ENTER]	The 10 random numbers are: 36 13 6 17 21 ~~21~~ 20 5 31 16 48 Not that repetitions can occur but 21 cannot be used twice.

C-12 Matrices

Given the matrices $A = \begin{bmatrix} 1 & -4 & 6 \\ 2 & -1 & 0 \\ 5 & -3 & 6 \end{bmatrix}$ and $B = \begin{bmatrix} 4 & 15 \\ -2 & 6 \\ 2 & 5 \end{bmatrix}$.

EXAMPLE 1: Find -2.5AB.
Solution:

Keystrokes	Explanation
[2nd] [MATRX] [F2] <EDIT> [A] [ENTER]	Choose matrix A and enter the dimensions.
[3] [ENTER] [3] [ENTER]	
[1] [ENTER] [2] [ENTER] [5] [ENTER] [(-)] [4] [ENTER] [(-)] [1] [ENTER] [(-)] [3] [ENTER]	Enter the matrix elements column by column.
[6] [ENTER] [0] [ENTER] [6] [EXIT]	Choose matrix B and enter the dimensions.
[F2] <EDIT> [B] [ENTER]	
[3] [ENTER] [2] [ENTER]	
[4] [ENTER] [(-)] [2] [ENTER] [2] [ENTER] [15] [ENTER] [6] [ENTER] [5]	Enter the matrix elements.
[EXIT] [EXIT]	Return to the Home Screen.
[(-)] [2.5] [2nd] [MATRX] [F1] <NAME> [F1] <A> [x] [F2] <B> [ENTER]	Enter the matrix operations. The result is $\begin{bmatrix} -60 & -52.5 \\ -25 & -60 \\ -95 & -217.5 \end{bmatrix}$.

EXAMPLE 2: Find A^{-1}.

Keystrokes	
2nd MATRX F1 <NAME> F1 <A> 2nd x^{-1} ENTER	The operations are done from the Home Screen. The result rounded to three decimal places is: $A = \begin{bmatrix} -.167 & .167 & .167 \\ -.333 & -.667 & .333 \\ -.028 & -.472 & .194 \end{bmatrix}$.

Some numbers may look like decimals. However upon examining the matrix using the arrow keys you will find they are decimals that are very close to zero. This situation can be eliminated by setting the number of decimal places to a number, say 3 (see Section C-3 of this manual).

EXAMPLE 3: Solve the system of equations using the inverse of the coefficient matrix. Round answers to two decimal places.

$$\begin{aligned} x - 4y + 6z &= 2.5 \\ 2x - y \quad\quad &= 3.2 \\ 5x - 3y + 6z &= -3.4 \end{aligned}$$

Solution: The inverse of the coefficient matrix was found above in Section C-12 Example 2. Enter the constants as matrix B and find $A^{-1}B$.

Keystrokes	
2nd MATRX	
F2 <EDIT> 2 :<B> ENTER	Enter the matrix dimensions and enter the matrix elements.
3 ENTER 1 ENTER 2.5 ENTER	
3.2 ENTER (-) 3.4 EXIT	Return to the Home Screen.
EXIT 2nd MATRX F1 <NAME>	Perform the matrix operations.
F1 <A> 2nd x^{-1} × F2 <B> ENTER	The result is $\begin{bmatrix} -.45 \\ -4.1 \\ -2.24 \end{bmatrix}$. The solution to the system of equations is $x = -.45$, $y = -4.1$, $z = -2.24$.
EXIT EXIT	Return to the home screen.

EXAMPLE 4: Find the solution to the system of Example 3 above using Gauss-Jordan reduction. In other words, find the reduced form of the matrix $\begin{bmatrix} 1 & -4 & 6 & 2.5 \\ 2 & -1 & 0 & 3.2 \\ 5 & -3 & 6 & -3.4 \end{bmatrix}$.

Keystrokes	
2nd MATRX F2 <EDIT>	Enter the matrix dimensions.
A ENTER 3 ENTER 4 ENTER	
1 ENTER (-) 4 ENTER 6	Enter the matrix elements.
ENTER 2.5 ENTER 2 ENTER	
(-) 1 ENTER 0 ENTER 3.2	
ENTER 5 ENTER (-) 3 ENTER	
6 ENTER (-) 3.4 EXIT EXIT	

2nd MATRX F4 <OPS>
MORE F5 <mRAdd>
(-) 2 , EXIT F1 <NAME>
F1 <A> , 1 , 2) ENTER
STO▸ F1 <A>

We wish to multiply by -2, matrix A, row 1 and add it to row 2. The result will be put in row 2.

```
mRadd(-2,A,1,2)
```

The result is $\begin{bmatrix} 1 & -4 & 6 & 2.5 \\ 0 & 7 & -12 & -1.8 \\ 5 & -3 & 6 & -3.4 \end{bmatrix}$.

Store the result in matrix A.

ENTER 2nd MATRX
F4 <OPS> MORE F5 <mRAdd>
(-) 5 , EXIT F1 <NAME>
F1 <A> , 1 , 3) ENTER

Now multiply row 1 by -5 and add it to row 3 putting the result in row 3.

```
mRadd(-5,A,1,3)
```

The result is $\begin{bmatrix} 1 & -4 & 6 & 2.5 \\ 0 & 7 & -12 & -1.8 \\ 0 & 17 & -24 & -15.9 \end{bmatrix}$.

STO▸ A ENTER

Store the result in matrix A.

2nd MATRX F4 <OPS> MORE
F4 <multR> 1 ÷ 7 , ALPHA A
, 2) ENTER
STO▸ A

Multiply row 2 by 1/7 .

The result is $\begin{bmatrix} 1 & -4 & 6 & 2.5 \\ 0 & 1 & -1.71 & -.26 \\ 0 & 17 & -24 & -15.9 \end{bmatrix}$

rounded to two decimal places.
Store the result in matrix A.

Continue in this manner until you get $\begin{bmatrix} 1 & 0 & 0 & -.45 \\ 0 & 1 & 0 & -4.1 \\ 0 & 0 & 1 & -2.24 \end{bmatrix}$. The solution to the system of equations is $x = -.45$, $y = -4.1$, $z = -2.24$.

> Note: To swap rows of a matrix use F4 <OPS> MORE F2 <rSwap> from the MATRX menu. The command rSwap(A,2,3) will swap rows 2 and 3 in matrix A.

C-13 Logarithmic, Exponential, and Hyperbolic Functions

EXAMPLE 1: Graph $f(x) = 10^{3.2x}$ for $-.5 < x < .5$.
Solution: Set the RANGE to [-.5,.5].1 by [-10,10]1.

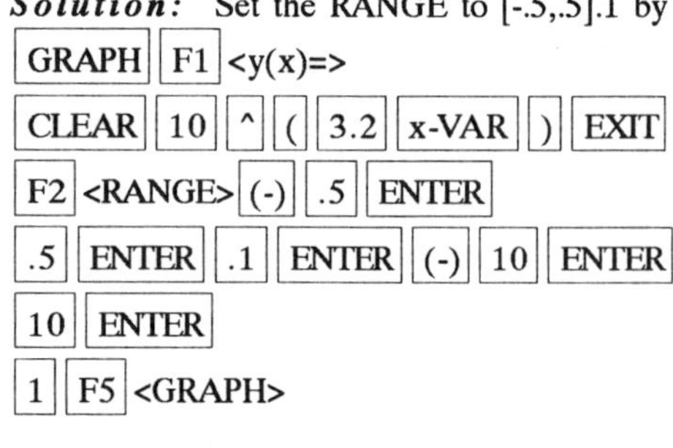

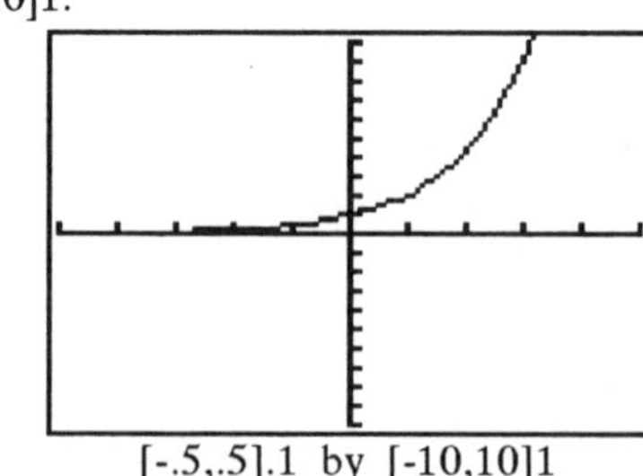

[-.5,.5].1 by [-10,10]1

EXAMPLE *2*: Graph $y = \dfrac{e^{x} + e^{-x}}{2}$.

Solution: This could be entered using [(] [2nd] [e^x] [x-VAR] [+] [2nd] [e^x] [(-)] [x-VAR] [)] [÷] [2] . However the built-in hyperbolic cosine function can also be used.

[GRAPH] [F1] <y(x)=> [CLEAR]
[2nd] [MATH] [F4] <HYP> [F2] <cosh>
[x-VAR] [EXIT] [EXIT] [F3] <ZOOM>
[F4] <ZSTD>

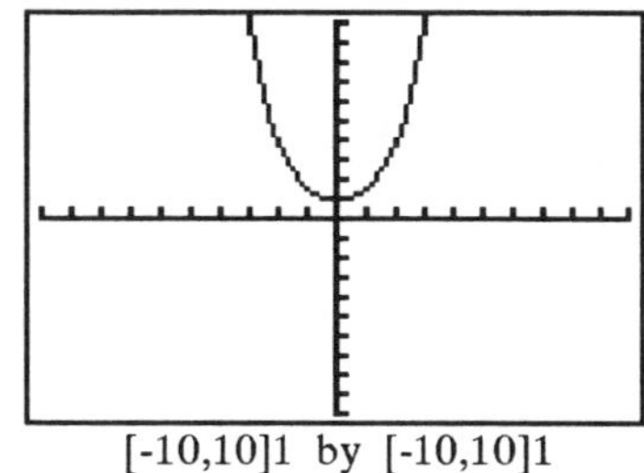

[-10,10]1 by [-10,10]1

EXAMPLE *3:* Solve $\ln(x^{2}+x) = 3x + 2$
Solution:

[GRAPH] [F1] <y(x)=> [CLEAR]
[LN] [(] [x-VAR] [^] [2]
[+] [x-VAR] [)] [ENTER]
[CLEAR] [3] [x-VAR] [+] [2]
[ENTER]
[EXIT] [F3] <ZOOM>
[F4] <ZSTD>

Enter the function $f(x) = \ln(x + x^{2})$ as Y1 in the function list.

Enter the function $g(x) = 3x + 2$ as Y2 in the function list.

Graph them on the same graphing screen. Use ZOOM and TRACE to find the x value of the point of intersection. This will be the solution to the equation.

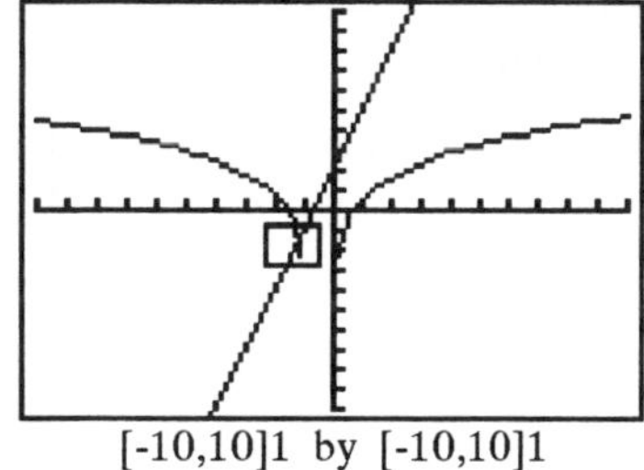

[-10,10]1 by [-10,10]1

The solution is -1.18, rounded to two decimal places.

C-14 Trigonometric Functions

EXAMPLE *1:* Evaluate $f(x) = \cos x + \csc x$ at (A) 53°45'18" (B) 78.52 radians

Solution: ***(A)***

[GRAPH] [F1] <y(x)=> [CLEAR]
[COS] [x-VAR] [+] [1] [÷] [SIN]
[x-VAR] [EXIT] [EXIT]

Store the function.

[2nd] [MODE] [▼] [▼] [▶] [ENTER]

Change the calculator to degree angle measure.

CLEAR 53 + 45 ÷ 60 + 18 ÷ 3600 STO▸ x-VAR ENTER — Change the angle to decimal degrees while storing it as the variable x.

2nd ALPHA y 1 ENTER — Evaluate the function. The result to two decimal places is 1.83.

Solution: (B)

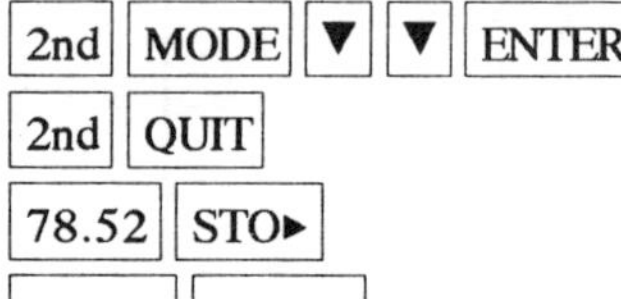

2nd MODE ▼ ▼ ENTER — Change the calculator to radian angle measure.

2nd QUIT — Since the function was stored in part (A) above, we need only to store the angle measure and evaluate the stored function.

78.52 STO▸ x-VAR ENTER — Store 78.52 as x.

2nd ALPHA y 1 ENTER — Recall the function. The result to two decimal places is 49.47.

EXAMPLE 2: Graph $f(x) = \cos x + \csc x$.

Solution: The function was entered in the calculator in Example 1 above.

GRAPH MORE F3 <FORMT> ▼ ▼ ▶ ENTER — Change the graph mode to Dot. This will eliminate the "vertical lines."

F5 <GRAPH> F3 <ZOOM> MORE F3 <ZTRIG> — This option automatically sets the graph RANGE to [-8.24,8.24]1.57 by [-4,4]1.

[-8.24,8.24]1.57 by [-4,4]1

C-15 Differentiation and Integration

EXAMPLE 1: Find the equation of the tangent line to $f(x) = \sqrt{x^5 - \sqrt[3]{x}}$ at $x = 3.54$. Use two decimal place accuracy.

Solution: The function and it's derivative can be evaluated at this value of x .

Keystrokes	
GRAPH F1 <y(x)=> CLEAR 2nd	Enter the function and store it as y1.
√ (x-VAR ^ 5 - x-VAR ^	
(1 ÷ 3))	
2nd QUIT 3.54 STO▸	Get the Home Screen and store the value of x .
x-VAR ENTER	
2nd VARS F1 <ALL>	Use the arrow keys to highlight the function.
▼ ... ▼ ENTER ENTER	Press ENTER to evaluate the function. $f(3.54)$ = 23.55
2nd CALC F3 <der1> 2nd alpha	Get the numeric derivative built-in function: `der1(y1,x,3.54)`
y 1 , x-VAR , 3.54)	This will evaluate the derivative with respect to x of the function stored in y1 at x = 3.54. The result is 16.67. (The default step size used in x increments is .001.)
ENTER	

The equation of the tangent line is y - 23.55 = 16.67(x - 3.54) .

EXAMPLE 2: Is the function $f(x)$ defined in Example 1 concave upwards at x = 3.54 ?

Solution:

Keystrokes	
Graph F1 <y(x)=> 2nd √ (	Enter the function and store it as y1.
x-VAR ^ 5 - x-VAR ^ (1	
÷ 3)) 2nd QUIT	
2nd CALC F4 <der2> 2nd alpha	Use the second derivative function. `der2(y1,x,3.54)`
y 1 , x-VAR , 3.54) ENTER	The result is positive. Hence the function is concave upwards at x = 3.54 .

EXAMPLE 3: Find $\int_{1.32}^{2.58} \sqrt{x^5 - \sqrt[3]{x}}\, dx$. Use two decimal place accuracy.

Solution:

Keystrokes	
Graph F1 <y(x)=> 2nd √ (x-VAR	Enter the function and store it.
^ 5 - x-VAR ^ (1 ÷ 3	
)) 2nd QUIT	
2nd CALC F5 <fnInt> 2nd alpha	Get the numeric integration function built into the calculator function: `fnInt(y1,x,1.32,2.58)`
y 1 , x-VAR , 1.32 , 2.58)	This will evaluate the definite integral . The result is 6.95 . The tolerance in calculating the definite integral is .0001.
ENTER	

C-16 Statistics

EXAMPLE 1: Given the following data find: (A) The mean. (B) The sample standard deviation . (C) Histogram with first class beginning at 10 and width 4. Use two decimal places. (D) Repeat Parts A and B after deleting the data value 89 from the list. How did eliminating this outlier effect the mean and standard deviation?

23	56	73	12	89	52	45
38	18	21	15	37	42	51
19	40	57	38	54	60	39

Solution (A) and (B):

STAT F2 <EDIT>

Get the statistics menu.

ENTER ENTER F5 <CLRxy>

Clear the statistical registers.

23 ENTER ENTER 38 ENTER
ENTER 19 ENTER ENTER Etc.

Enter the data in the x1 position. The y1 position is used for frequency if there is more than one piece of data with the same number. Enter all data. There are 21 values.

If you make a mistake, use the arrow keys to highlight the value and type it in again.

To remove a data value, use the arrow keys to get the *x* data value and press F1 <DELi>.

EXIT

Exit the data entry screen.

F1 <CALC> ENTER ENTER
F1 <1-VAR>

The mean is 41.86 and the sample standard deviation is 19.82.

Solution (C):

EXIT EXIT GRAPH F1 <y(x)=>
F5 <SELCT>

This will deselect the function the cursor is on. The = sign will no longer be highlighted. Deselect all functions stored or delete them using CLEAR.

EXIT F2 <RANGE>
0 ENTER 100 ENTER 4 ENTER
(-) 5 ENTER 5 ENTER 1
EXIT

The smallest value is 12 so let xMin=0. The greatest value is 89. Let us set xMax=100. Let's use xScl=4. This sets the histogram class width.
Let yMin=-5 to get some room below the histogram. We do not know the frequencies yet so let yMax=5 for a first try. Let yScl=1.

STAT F3 <DRAW> F5 <CLDRW>

Clear all drawings from the display screen.

F1 <HIST>

Draw the histogram.

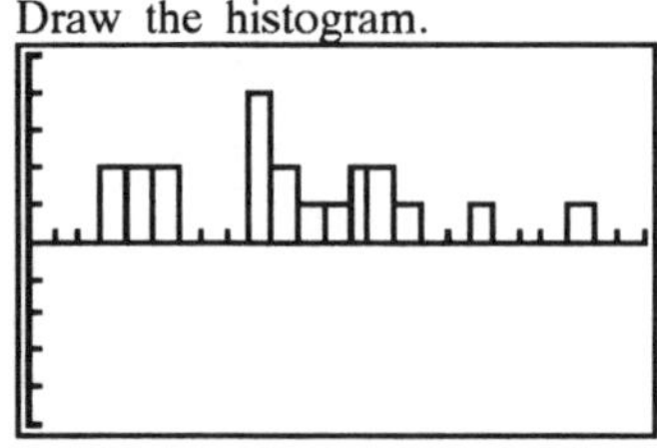

Solution (D):

Keystrokes	Explanation
EXIT F2 <EDIT> ENTER ENTER	Get the data list and arrow down to the 89.
▼ ... ▼ F2 <DELi>	Delete the item.
EXIT F1 <CALC> ENTER ENTER F1 <1-VAR>	Get the statistics menu and the one variable sample statistics. The mean is now 39.5 and the sample standard deviation is 17.05. Notice how much influence an outlier has!

EXAMPLE 2: Given the following bivariate data. (A) Find the correlation coefficient. (B) Find the least squares line. (C) Graph the least squares line. (D) Plot the scatter diagram . (E) Sort the data on x and join the points using straight line segments.

x	5	8	10	29	9	24
y	3	4	8	20	3	19

Solution (A) & (B):

Keystrokes	Explanation
STAT F2 <EDIT>	Get the statistics menu.
ENTER ENTER F5 <CLRxy> 5 ENTER 3 ENTER 8 ENTER 4 ENTER 10 ENTER 8 ENTER etc. EXIT	Clear the statistical registers. Enter the data.
F1 <CALC> ENTER ENTER F2 <LINR>	Linear regression. r = .98 a=-1.72 and b=.79 So the least squares line is $\hat{y}$ = .79x - 1.72 .

Solution (C):

Keystrokes	Explanation
EXIT EXIT	
GRAPH F2 <RANGE> 0 ENTER 30 ENTER 1 (-) 10 ENTER 30 ENTER 1	Set the graph RANGE. Observe the data to determine the RANGE values. Set the y value lower so if you trace, the coordinates at the bottom of the screen do not cover up the graph.
F1 <y(x)=> F5 <SELCT> EXIT EXIT	Deselect (or clear) all functions on the function list. See Example 1 above.
STAT F3 <DRAW> F4 <DRREG>	Draw the regression line that you just calculated.

Solution (D):

Keystrokes	Explanation
F2 <SCAT>	The scatter diagram will draw on the same graph as the least squares line.

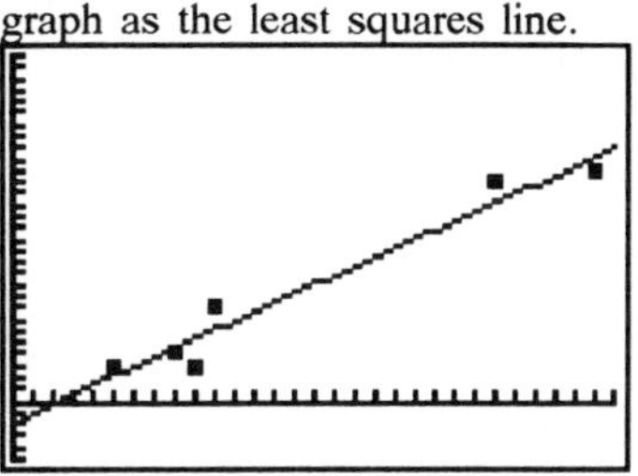

Solution (E):

EXIT F2 <EDIT> — Sort the data on *x* variable.

ENTER ENTER F3 <SORTX>

EXIT F3 <DRAW> F5 <CLDRW> — Clear the drawing. Draw the scatter diagram.

F2 <SCAT> F3 <xyLINE> — Draw the line joining the points. They will join in the order of the *x* data in the data list.

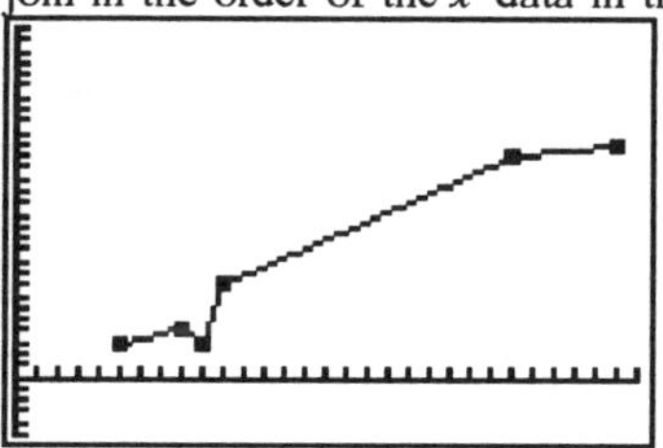

C-17 Setting the Display Mode for Numbers

Numbers can be entered into the calculator in scientific notation, a set number of significant digits, or a fixed number of decimal places.

EXAMPLE 1: Calculate $(9.35417\times 10^4)(-3.64389\times10^{-6})$ using:
(A) scientific notation (B) two significant digits (C) four decimal place accuracy.

Solution (A):

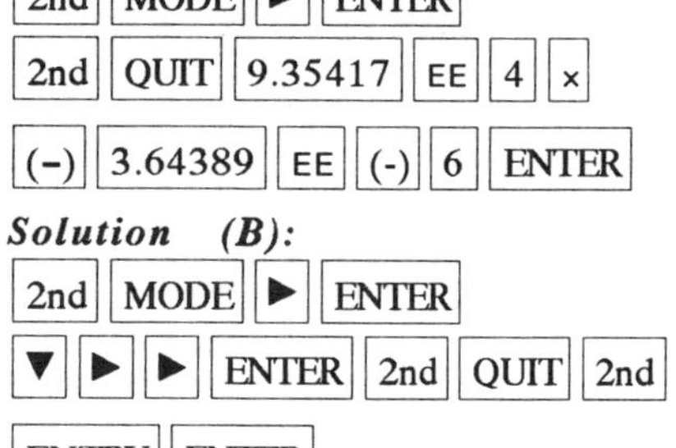

2nd MODE ► ENTER

2nd QUIT 9.35417 EE 4 ×

(-) 3.64389 EE (-) 6 ENTER

The calculator is set to scientific mode with floating decimals. The result displayed on the calculator is -3.40855665213E-1.

Solution (B):

2nd MODE ► ENTER

▼ ► ► ENTER 2nd QUIT 2nd

ENTRY ENTER

Set the calculator to scientific notation with 1 digit instead of floating point. Return to the Home Screen.

Use the replay key to replay the calculation just done in Solution (A). The result to two significant digits is -3.4E-1 .

Solution (C):

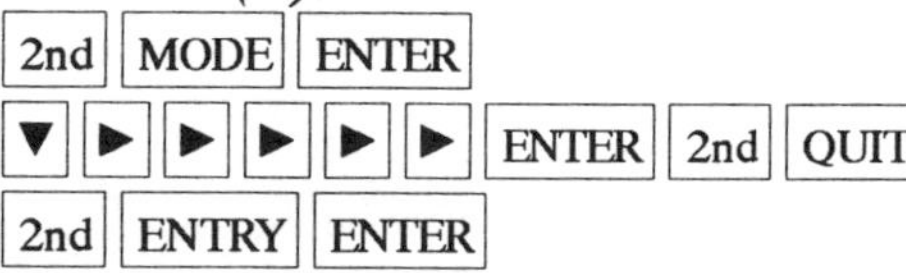

2nd MODE ENTER — Set the calculator to normal mode with 4 digits. Return to the Home Screen.

▼ ► ► ► ► ► ENTER 2nd QUIT

2nd ENTRY ENTER — Use the replay key to replay the last calculation. The result to four decimal places is -.3409.

Change the calculator back to floating point. The examples in this manual are done using this setting unless otherwise noted.

Answers to Selected Problems

Chapter 1

3. -5.25, -5.09, -5.0099, -5.000999. The value of $Q(x)$ gets close to -5 as x gets close to 2.

5.

h	.2	.1	.01	.001	→	0
$\frac{f(1.5+h) - f(1.5)}{h}$	.9	.8	.71	.701	→	.7

7.

x	$p(x)$	$p(x)$-p
2	1118	8
5	1010	20
10	830	15
14	686	9
21	434	1

11. (E) The slope of the line is greater but the y intercept remains the same. Hence the x intercept of $g(x) = 2.2x - 4.2$ would be less than that of $f(x) = 1.2x - 4.2$. In fact, the x intercept of $g(x)$ is 1.9.

13. Yes.

15. The two lines are almost parallel. Compare the slopes 5.23/8.92 and 5.21/8.89.

17. The second graph is stretched vertically . All y values of the second graph are three times those of the first graph.

The third graph is stretched vertically by a factor of 1.5 and reflected about the x axis. All y values of the first graph are multiplied by -1.5 to obtain the third graph.

19. (A) & (B)

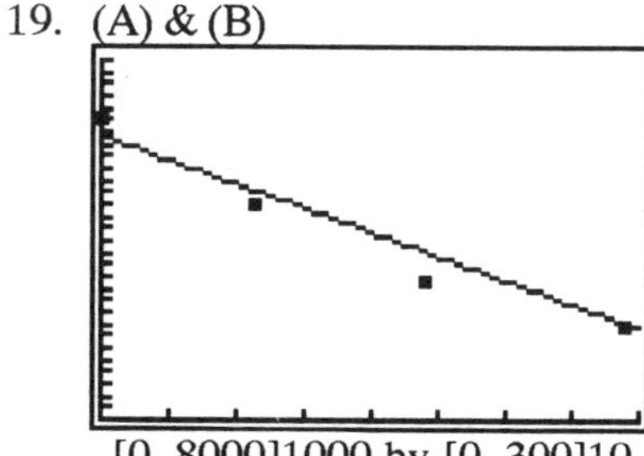

[0, 8000]1000 by [0, 300]10

(C) Use trace to find the y intercept of 237.92 rounded to two decimal places. Hence one point on the line is (0, 237.92).

Use trace to find the coordinates of the point having x coordinate of 8000.

This point is (8000, 62.52). Use these two points to find the slope of the line. The slope is -.02. Hence the equation of the line with coefficients rounded to two decimal places is $y = 237.92 - .02x$.

(D) 161.18

21. (A $C = 360{,}000 - 900p$
(B) $R = xp = (9000 - 30p)p$
(C)

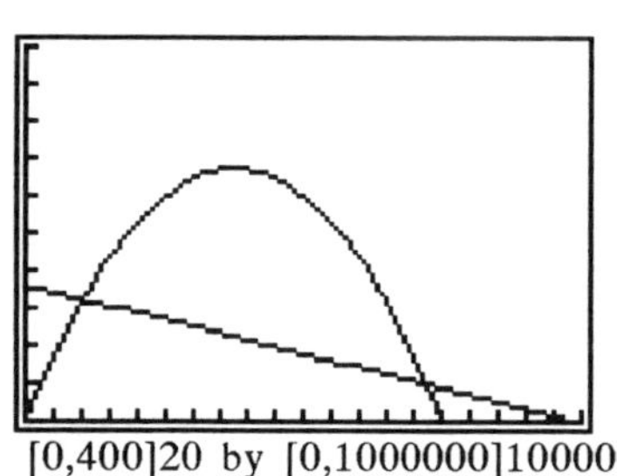

[0,400]20 by [0,1000000]100000

(C) Profit when $42 < x < 288$. Loss when $0 < x < 42$ or $288 < x < 300$.

(D) \$42, \$288 (E) \$150

Chapter 2

13. 1.07

15. $x < -1.41 \cup .21 < x$.

17.

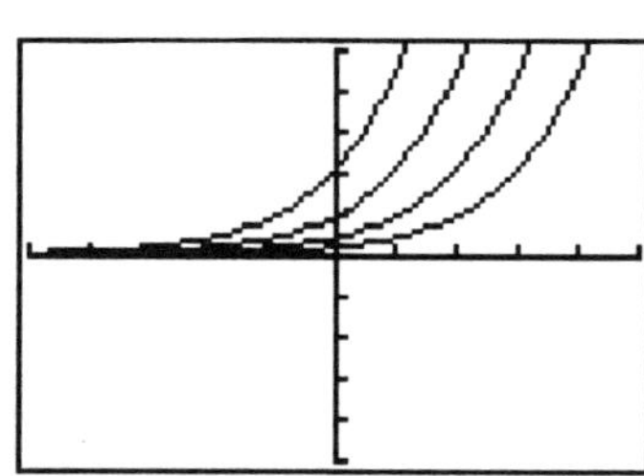

[-5,5]1 by [-5,5]1

Adding a constant to the x shifts the graph to the left.
Subtracting a constant to the x shifts the graph to the right.

19. 22.52825

21.

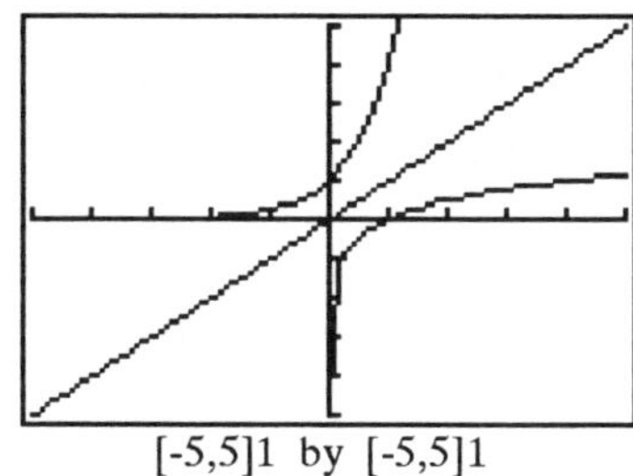

[-5,5]1 by [-5,5]1

Use the change-of-base formula to graph the inverse function.

Chapter 3

3. 2, 175.70, 4351.41, 10878.52

9.

Quarter Number	Principal at Start of Quarter	Amount at End of Quarter	Interest Earned During the Quarter
1	6925.00	7037.53	112.53
2	7037.53	7151.89	114.36
3	7151.89	7268.11	116.22
4	7268.11	7386.22	118.11
5	7386.22	7506.24	120.03
6	7506.24	7628.22	121.98

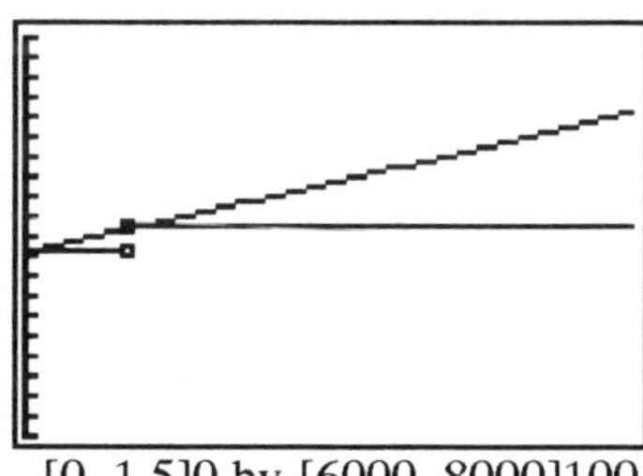

[0, 1.5]0 by [6000, 8000]100

15. S = \$1256.85, A = \$1318.52; Total interest earned = \$75.05.

17.

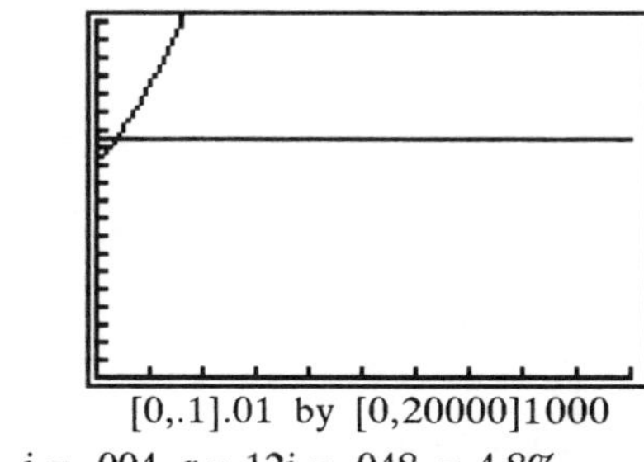

[0,.1].01 by [0,20000]1000

[0,.01].001 by [0,20000]1000

$i = .004$, $r = 12i = .048 = 4.8\%$.

23. The payment is \$307.38. The total interest paid is \$188.56.

Period Number	Payment	Interest	Balance Reduction	Unpaid Balance
1	307.38	28.58	278.80	3221.20
2	307.38	26.31	281.07	2940.13
3	307.38	24.01	283.37	2656.76
4	307.38	21.70	285.68	2371.08
5	307.38	19.36	288.02	2083.06
6	307.38	17.01	290.37	1792.69
7	307.38	14.64	292.74	1499.95
8	307.38	12.25	295.13	1204.82
9	307.38	9.84	297.54	907.28
10	307.38	7.41	299.97	607.31
11	307.38	4.96	302.42	304.89
12	307.38	2.49	304.89	0.00

Chapter 4

3. The lines appear to intersect in a single point on the graph using [-10,10]1 by [-10,10]1. Graph using viewing window variables of [.4,.5].01 by [-2.5,-2.0].1 to see that the lines do not intersect in a single point.

5. There will still be no solution. Changing the constant to 14 shifts the line downward 2 units.

11. $x = .29$, $y = -2.85$, $z = -3.65$

21. $x = 1.04$, $y = -.95$, $z = 1.07$

23. $\begin{bmatrix} \$54 & \$48 \\ \$36 & \$32 \end{bmatrix}$ and $\begin{bmatrix} \$60.53 & \$48 \\ \$40.35 & \$32 \end{bmatrix}$

The labor costs increase \$6.53 per ski for the trick ski and \$4.35 per ski for the slalom ski.

Chapter 5

5. Corner points are (0, 7.14), (0, 10.06), (6.13, 0), (26.83, 0), and (2.66, 1.45)

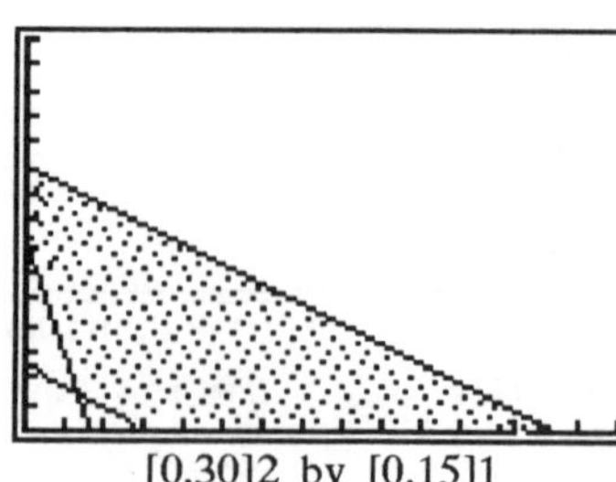

[0,30]2 by [0,15]1

13. Maximum at (26.83, 0), minimum at (0, 7.14).

15. A maximum profit of \$2520 is achieved if 28 standard tents and 0 expedition tents are manufactured and sold. Substituting the solution into the slack variable equations shows that $s_1 = 4$ and s_2 is zero. This means that there are 4 hours in the cutting department that are not used. All available hours in the assembly department are used.

Chapter 6

3. 2,496,960; 6,991,488; 621,075; 1.366,365

5. Using a random number generator seed of 10 and entering [IPart] (25 x [RAND]) we get: 11, 13, 6, 17, and 21. Note that we use 25 as the number in the command since we want to generate numbers x such that $0 \le x \le 24$.

7. (A) $\begin{bmatrix} .294 & .130 & .015 \\ .106 & .083 & .092 \\ .076 & .065 & .139 \end{bmatrix}$ (B) .618

(C) P(H) =.476; the events are not independent.

11. (A) (B) .25

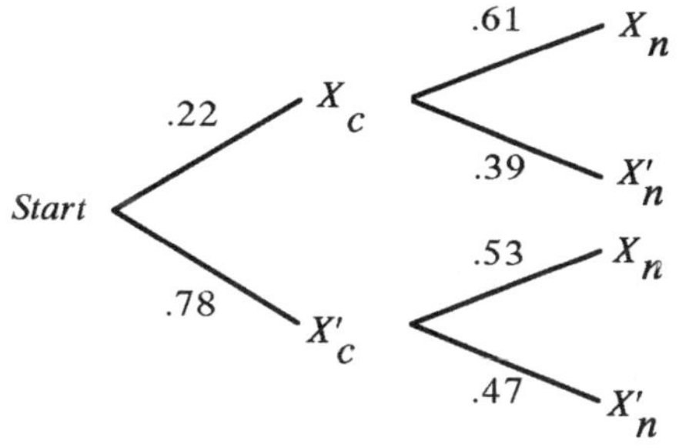

15. (A) .1

(B) In the long run this player will win \$.10 each game on the average. Numbers 5 and 6 are both the best numbers for the player to choose since they have the highest probability.

17. -2.11

Chapter 7

7.

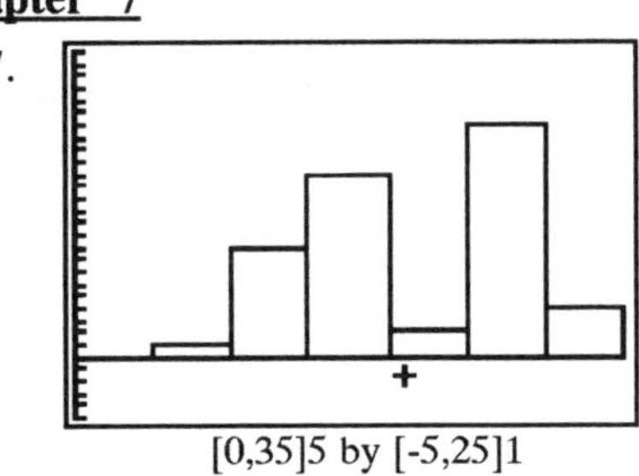

[0,35]5 by [-5,25]1

11. $\bar{x} = 21.12$, s = 7.07.

13. P(47)+ P(46)+ P(45) = .422 chance that 3, 4, or 5 will not recover.

Chapter 8

5. (A)

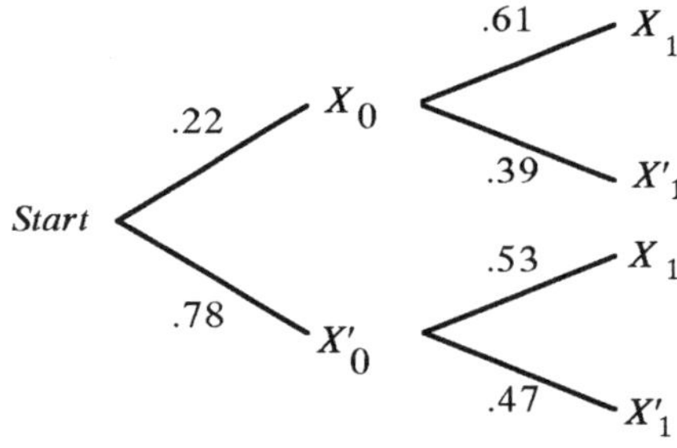

(B) Use Bayes formula.

$$P(X_0 \mid X_1) = \frac{P(X_0 \text{ and } X_1)}{P(X_1)}$$

$$= \frac{P(X_0 \text{ and } X_1)}{P(X_0 \text{ and } X_1) + P(X_0' \text{ and } X_1)}$$

$$= \frac{.22x.61}{.22x.61 + .78x.53} = .245$$

7. Transition matrix is

$$\begin{array}{c} \\ X_0 \\ X_0' \end{array} \begin{array}{c} X_1 \quad X_1' \\ \begin{bmatrix} .61 & .39 \\ .53 & .47 \end{bmatrix} \end{array}.$$

The initial state is [.22 .78].

(A) After five purchases, the company will have 58% of the market.
(B) In the long run, the company will have 58% of the market.

9. (A) After five purchases, Product X will have 35% of the market and Product Y will have 29% of the market.
   (B) In the long run, Product X will have 35% of the market and Product Y will have 30% of the market.

Chapter 9

7.

x	.9	.99	.999	→1←	1.001	1.010	1.100
$f(x)$	32	302	3002		-2998	-298	-28

The limit does not exist. Hence the function is not continuous.

9. Not continuous at $x = 2$. Yes, it is the same function as in Example 2.

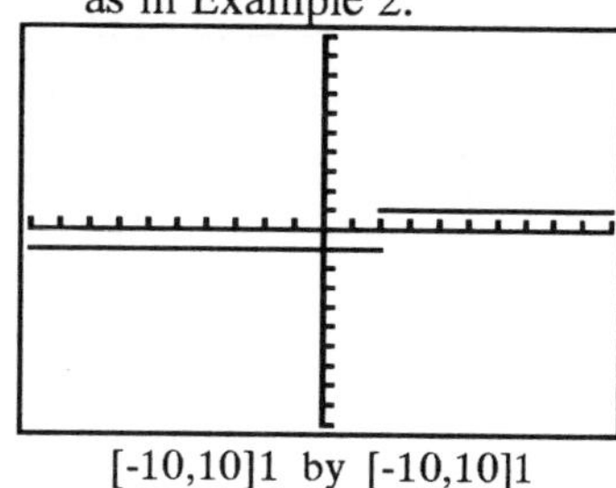

[-10,10]1 by [-10,10]1

17. A point of discontinuity at x=-.5

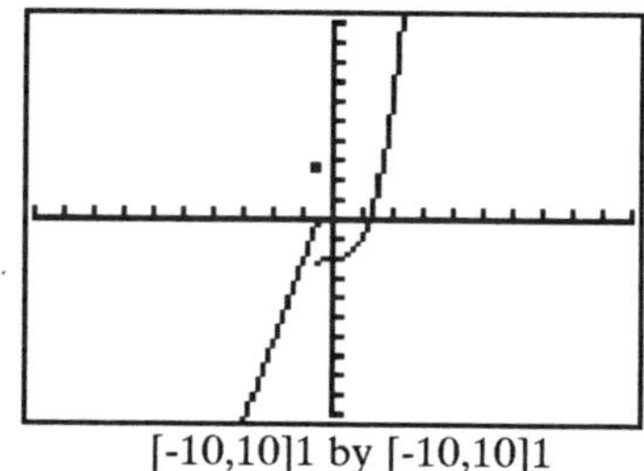

[-10,10]1 by [-10,10]1

(The point (-.5, 3.25) may not be graphed on your calculator.)

19. Graph and use trace to see that the y values do not approach a number as x gets very large.

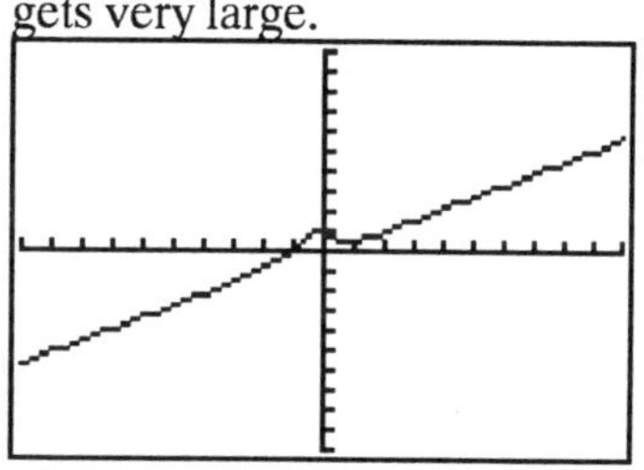

[-10,10]1 by [-10,10]1

Change the viewing window variables to [0,1000]100 by [0,1000]100 to see the right-hand tail.
Then change the viewing window variables to [-1000,0]100 by [-1000,1000]100 to see the left-hand tail.

21. Viewing window variables: [-1,5]1 by [-1,3]1

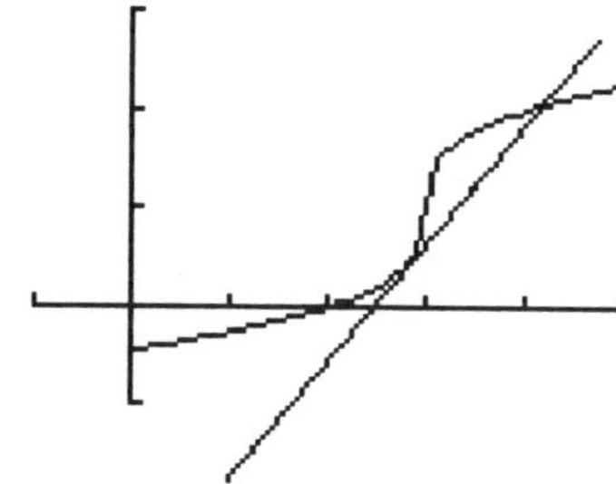

h	.01	.005	.001	.0001	.00001→ 0
d.q.	21.54	34.20	100	464.2	21.54 → limit does not exist.

d.q. = difference quotient at $x = 3$.

Chapter 10

7.

x	.9	.99	.999	→1←	1.001	1.010	1.100
$f(x)$	32	302	3002		-2998	-298	-28

The limit does not exist. Hence the function is not continuous.

9. Not continuous at $x = 2$. Yes, it is the same function as in Example 2.

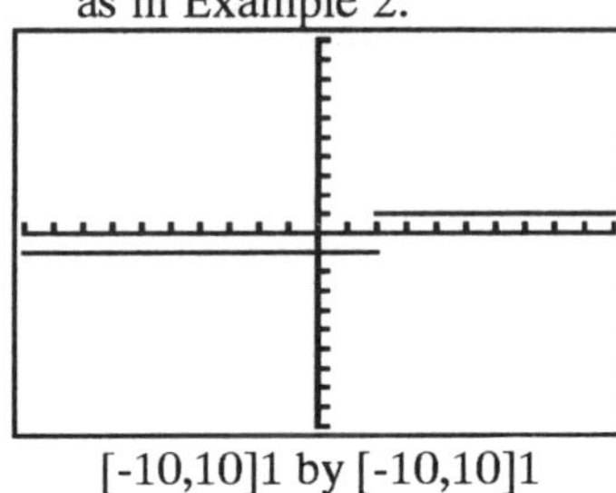

[-10,10]1 by [-10,10]1

11. A point of discontinuity at x=-.5

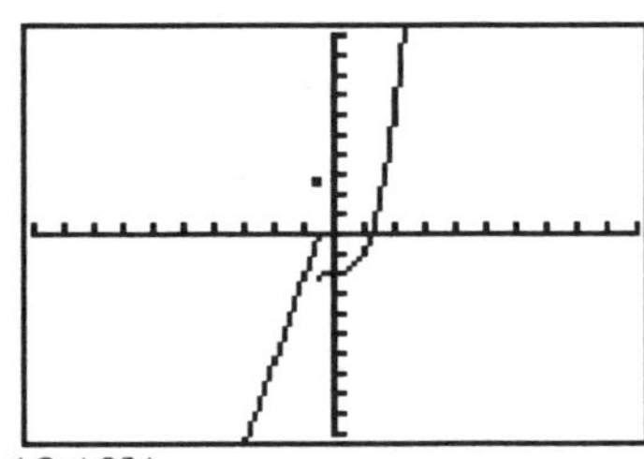

[-10,10]1 by [-10,10]1

(The point may not be displayed on your graphing screen.)

27. (A)

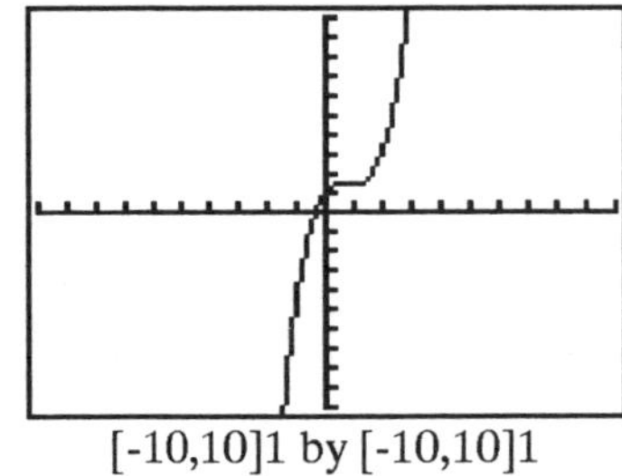

[-10,10]1 by [-10,10]1

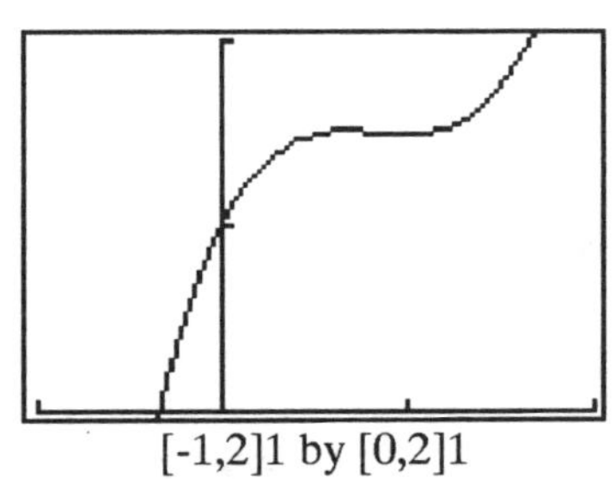

[-1,2]1 by [0,2]1

(B)

x	0	.4	.6	.8	1.0	1.2
$f(x)$	1.00	1.46	1.52	1.51	1.50	1.53
$f'(x)$	2.00	.48	.08	-.08	0.00	.32
$f''(x)$	-5.00	-2.60	-1.40	-.20	1.00	2.20

(C) x intercept = -.34
y intercept = 1

(G) (.83, 1.51)

(D) maximum at .67, minimum at 1

(E) $x < .67$, $1 < x$ or $(-\infty,.67)\cup(1,\infty)$

(F) $.67 < x < 1$ or $(.67, 1)$

(H) $x > .83$

(I) $x < .83$

(J) Look at the table. $f'(x)$ changes from negative to positive when x changes from .8 to 1.0

29. Graph and use trace to see that the y values do not approach a number as x gets very large.

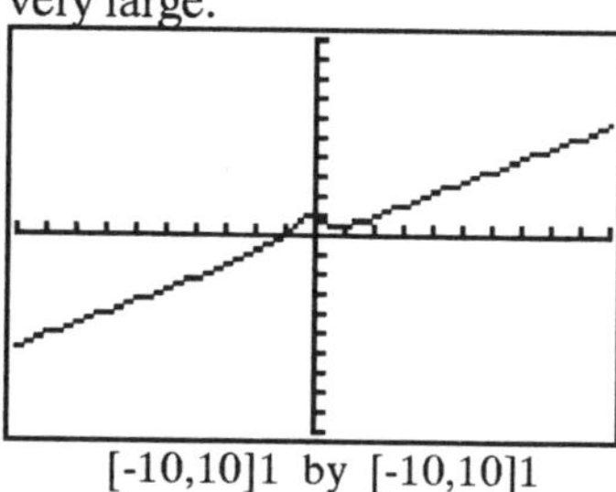

[-10,10]1 by [-10,10]1

Change the viewing window variables to [0,1000]100 by [0,1000]100 to see the right-hand tail.

Then change the viewing window variables to [-1000,0]100 by [-1000,1000]100 to see the left-hand tail.

31. Vertical asymptotes at $x = \pm\sqrt{5}$ (about ±2.24) and horizontal asymptote at $y = 2$.

33. Enter as 1.5 (LN (X - e^ (.03 X))) ÷ (LN 2) + 0.05 X + 2 .

x intercepts at 1.43 and 171.48. No y intercept. Vertical asymptotes at x=1.03 (find this algebraically by setting the argument of the ln function equal to zero) and x=171.48.

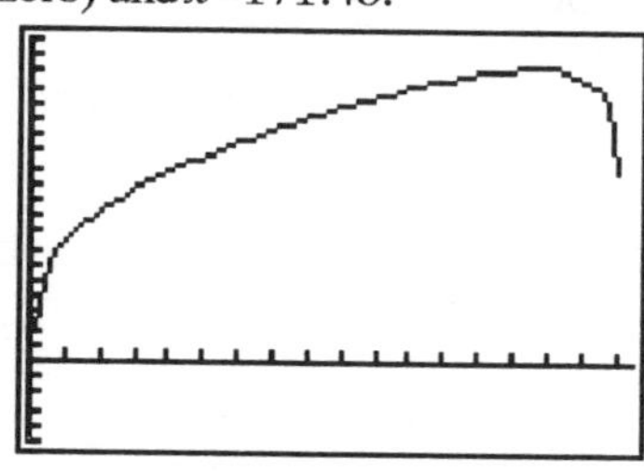

[0,175]10 by [-5,20]1

No horizontal asymptote.

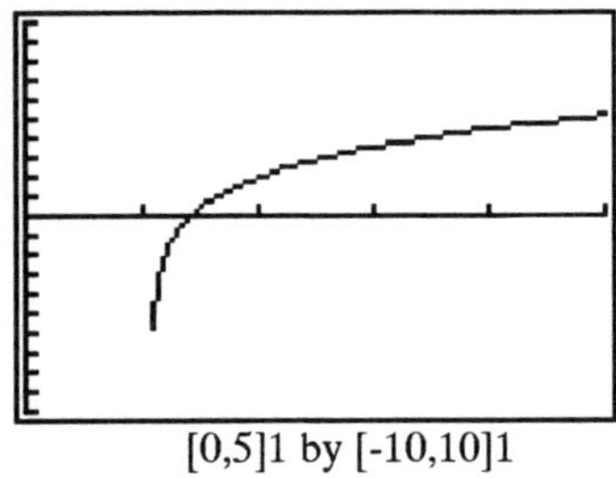

[0,5]1 by [-10,10]1

Graph using [171.4, 171.5].01
by [-5, 20]1 to see the other vertical asymptote.

Maximum when $x \approx 147.3$, $y \approx 18.4$.
No minimum.

Chapter 12

9. 17100 years.
15. 203.30
17. Δx=.6 The midpoints of the intervals are -.9, -.3, .3, .9 .
    Approximation is -.73 .

Chapter 13

5. 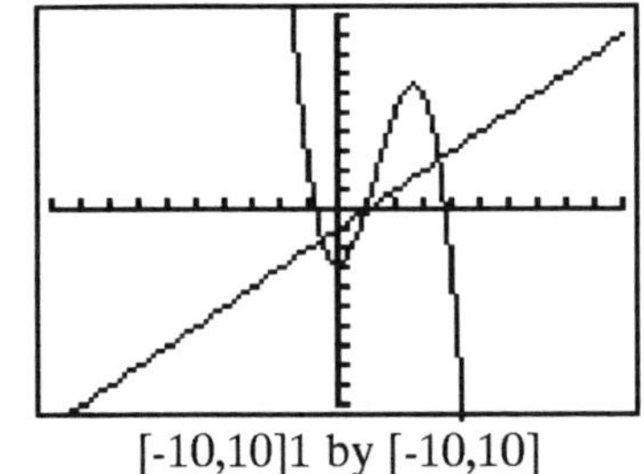

[-10,10]1 by [-10,10]
There three intersection points:
(-.56, -1.56), (1, 0), (3.56, 2.56)

Two integrals:

$$\int_{-.56}^{1} (x-1) - (-x^3+4x^2-3)\, dx \; +$$

$$\int_{1}^{3.56} (-x^3+4x^2-3) - (x-1)\, dx$$

The solution is 2.12+ 7.96 = 10.08

Chapter 14

3. (A) 310 units.
(B) The graph shows the number of units as the dependent variable and the thousands of dollars spent on labor each month as the independent variable when the amount spent on materials is held at \$50,000 per month.

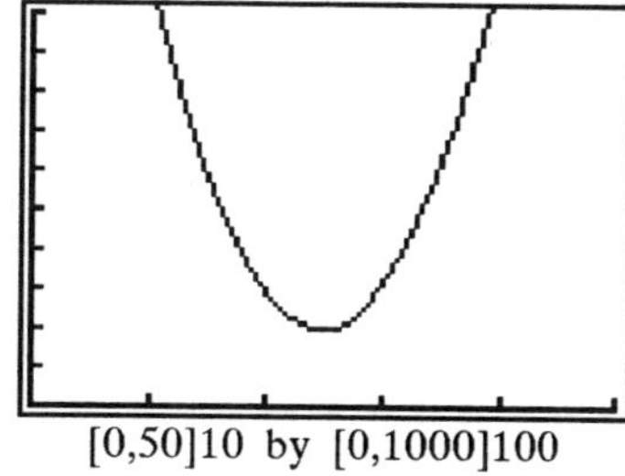

[0,50]10 by [0,1000]100

9. (A) 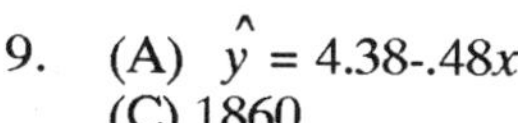 $\hat{y} = 4.38 - .48x$
(C) 1860
(D) \$6.00

(B)

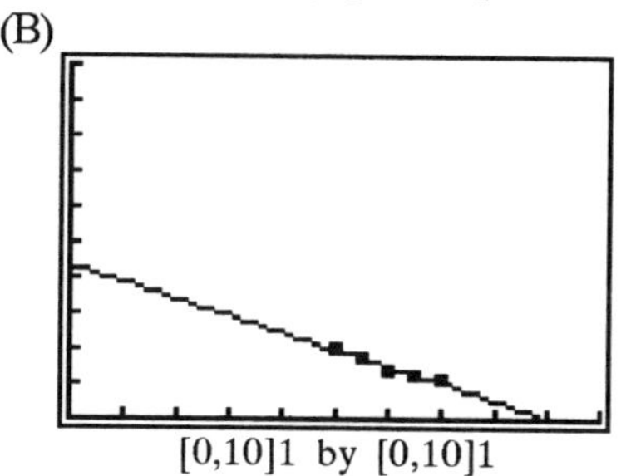

[0,10]1 by [0,10]1